Little toad, little toad

두껍아 두껍아 Dukkeoba, dukkeoba

I will give you an old house

헌집줄게 Heon jip julge

In return for a new house

새집다오   Sae jip dao

                                        Little toad, little toad

               두껍아 두껍아   Dukkeoba, dukkeoba

Go fetch some water

                            물 길어 오너라   Mul gireo oneora

                    I will build your house

     너희 집 지어 줄게   Neohui jip jieo julge

두껍아 두껍아   Dukkeoba, dukkeoba

Little toad, little toad
　　　　　　　　　　　　　　　두껍아 두껍아   Dukkeoba, dukkeoba
Your house is on fire
　　너희 집에 불났다 Neohui jibe bullatda
　　　　　　　　　　　Wend your way here with a pitchfork
쇠고랑 가지고 뚤레뚤레 오너라   Soegorang gajigo ttullettulle oneora

LITTLE TOAD, LITTLE TOAD

19. Mostra Internazionale di Architettura
La Biennale di Venezia
Partecipazioni Nazionali

UNBUILDING PAVILION

2025년 베니스비엔날레 제19회 국제건축전
한국관 전시도록 한국문화예술위원회 위원장 인사말

베니스비엔날레 한국관은 1995년 베니스시 카스텔로 공원 내 26번째 국가관으로 건립된 이후 세계에 한국 미술과 건축의 위상을 알리는 주요 무대로 자리매김하였습니다. 특히, 올해는 베니스비엔날레 한국관 개관 30년을 맞이하는 해로 그 의미가 더욱 각별합니다.

이런 유의미한 시점에 2025년 제19회 국제건축전 한국관 전시 «두껍아 두껍아: 집의 시간»을 선보이게 되어 매우 기쁩니다. 이번 전시는 한국관 건립 30주년을 계기로, 한국관의 초기 건립 과정을 되돌아보고, 한국관의 고유한 건축적 의미와 지속가능한 비전을 살펴보는 전시입니다. 한국관이 가지고 있는 두 개의 상반된 요소인 '임시성'과 '거주성'에 대해 두꺼비라는 화자의 시선과 함께 건축적 가치에 대해 질문하고, 베니스비엔날레가 맞이할 새로운 미래와 변화를 상상하는 자리가 되길 바랍니다.

2025년 한국관 전시는 세 명의 공동예술감독이 준비하였습니다. 전시 개최를 위해 애써주신 정다영, 김희정, 정성규 세 분의 예술감독과 전시 작품 제작을 위해 오랜 기간 수고하신 참여 작가들, 그리고 보이지 않는 곳에서 고생하신 전시추진단에게 감사의 말씀 전합니다. 또한, 올해 전시에 힘을 실어 주신 이케아 코리아, 삼성문화재단, 주성디자인랩|주|, 정림건축종합건축사사무소, (주)피앤앨/김석우, (주)해안종합건축사사무소, (주)간삼건축종합건축사사무소, (주)공간종합건축사사무소, (주)더시스템랩건축사사무소, 두오모, (주)제이아키브, LG 올레드 AI, LG 스탠바이미, VOLA, 러쉬코리아, 조병수건축연구소, 원오원아키텍츠, 하퍼스 바자 코리아, WOOYOUNGMI, 스트락스/어퍼하우스, 루나앤컴퍼니, 경민산업, Helinox, 한솔제지(주), 서울도시건축비엔날레 등 후원사와 협력 기관에도 깊은 감사의 인사를 드립니다.

한국문화예술위원회 위원장
정병국

Preface to the Korean Pavilion Catalog of
the 19th Venice Architecture Biennale

Since its establishment in 1995 as the 26th national pavilion in the Giardini di Castello of Venice, the Korean Pavilion has become a significant platform to showcase the excellence of Korean architecture and art on the international stage. This year holds even greater significance as it marks the 30th anniversary of the Pavilion's opening.

Against he background of this meaningful moment, I am delighted to introduce the Korean Pavilion's exhibition of the 19th International Architecture Exhibition of the Venice Biennale, *Little Toad, Little Toad: Unbuilding Pavilion*. This exhibition reflects on the 30-year history of the Korean Pavilion and re-examines the vision of sustainability and the architectural presence of the national pavilion. By discovering the interplay between temporariness and livability of the Korean Pavilion through the gaze of the "little toad," the exhibition inquires about architectural value and aesthetics, and envisions new futures and transformations of the Venice Biennale.

The exhibition of this year's Korean Pavilion was curated by the three co-artistic directors, Chung Dahyoung, Kim Heejung, and Jung Sungkyu. I would like to express my sincere appreciation to the co-artistic directors, assistant curator, and participating artists for their tireless effort in this project. Also, our appreciation goes out to the sponsors, including IKEA Korea, Samsung Foundation of Culture, JOOSUNG DESIGNLAB, JUNGLIM ARCHITECTURE, PNL Co., Ltd. / KIM, SEOK WOO, HAEAHN Architecture, Gansam Co.,Ltd, SPACE GROUP, THE_SYSTEM LAB, DUOMO, J.archiv, LG OLED AI, LG StanbyME, VOLA, LUSH, BCHO Architects Associates, ONE O ONE architects, Harper's BAZAAR Korea, WOOYOUNGMI, STRX/UPPERHOUSE, Luna&Company, KM Beam, Helinox, Hansol Paper, and Seoul Biennale of Architecture and Urbanism. Thank you again for everyone's invaluable support in bringing about this year's exhibition.

Byoung Gug Choung
Chairperson, Arts Council Korea

CONTENTS

7 PROLOGUE — FROM UNBUILING TO LIVING: RETHINKING THE PAVILION ··· Chung Dahyoung
13 VISUAL CURATING ESSAY ··· Kwak Seung-Chan

WORKS

106 EXHIBITION DRAWINGS
117 * OVERWRITING, OVERRIDING ··· Lee Dammy
131 * 30 MILLION YEARS UNDER THE PAVILION ··· Young Yena
147 * TIME FOR TREES ··· Heechan Park
163 * NEW VOYAGE ··· Kim Hyunjong

RESEARCH ESSAYS

179 THE HOUSE OF TREES: THE GIARDINI AND THE NATIONAL PAVILIONS ··· Kim Heejung, Jung Sungkyu
185 THE ESTABLISHMENT OF THE KOREAN PAVILION AT THE VENICE BIENNALE: BACKGROUND AND PROCESS, A STUDY THROUGH THE FRANCO MANCUSO DONATED ARCHIVES ··· Chun Jinyoung
193 WHEN WAS "THE LAST PAVILION"? ··· Alice S. Kim
206 SUSTAINABILITY, ARCHITECTURE AND THE VENICE BIENNALE ··· Song Ryul, Christian Schweitzer
218 TEMPORARY PERMANENT: A GENEALOGY OF REAL-SCALE ARCHITECTURE EXHIBITIONS ··· Léa-Catherine Szacka

223 A CURATORIAL FABLE ··· Jung Jinho
260 BIOGRAPHIES & COLOPHON

PROLOGUE
FROM UNBUILING TO LIVING:
RETHINKING THE PAVILION

CHUNG DAHYOUNG

The Korean Pavilion and the Toad

The Korean Pavilion exhibition at the 19th Venice Biennale International Architecture Exhibition explores the pavilion itself as its subject. As the Korean Pavilion marks its 30th anniversary this year, the exhibition aims to reinterpret its history while contemplating the pavilion's future and the sustainability of Venice Biennale pavilions. The 30-year milestone represents a crucial turning point in a building's life cycle. It marks the transition of a generation and signals a shift in architecture's social meaning. In Korea particularly, a building's 30th year often initiates discussions of demolition and reconstruction, marking the threshold between architectural life and death. Against this temporal backdrop, the exhibition *Little Toad, Little Toad* imagines the Korean Pavilion's future life through the act of reading and rewriting the pavilion itself. To this end, we view the pavilion not as a neutral, pristine exhibition space, but as an organic entity carrying multilayered meanings. We examine the pavilion as a "house of exhibitions" where memories and narratives accumulate.

The lyrics of 'Little Toad, Little Toad' serves as a metaphorical framework to unfold these ideas. This famous Korean children's folk song accompanies a mud play activity that evokes the grotto, which can be considered a primitive form of folly. The lyrics continue as children repeatedly build and destroy houses by piling up soil on the back of their hand by a stream or in a sandpit. The verse "I will give you an old house, In return for a new house" carries the meaning of regeneration through the exchange of old and new. The exhibition connects the Korean Pavilion's past and future through this framework of 'old house' and 'new house.' This opening line encapsulates the exhibition's aim to unbuild the inherent temporariness of pavilion architecture and reposition it as a living space, a "house." The verse "Your house is on fire" serves as a message of crisis and warning, calling for reflection on the social and environmental issues facing the Korean Pavilion. This message, which signals geopolitical changes among national pavilions in the Giardini, also alludes to Venice's recent predicament as one of the cities most vulnerable to the climate crisis. These lyrics have guided us toward exploring the sustainability of both the Korean Pavilion and the Venice Biennale.

The toad, the exhibition's metaphorical narrator, is a fabled creature symbolizing regeneration in both Eastern and Western cultures. In Korean tradition, the toad appears as a guardian spirit of the house and a harbinger of change, while in Western culture, it emerges as a symbol of transformation and alchemy. In the verse "Wend your way here with a pitchfork," the toad emerges as an agent of action, actively responding to crisis. The toad becomes the exhibition's invisible narrator, embodying the role of the architect.

The Changing National Pavilions

The national pavilions in the Giardini of Venice, which originated as temporary structures based on provisional agreements with Venetian authorities, now hold semi-permanent symbolic status. For well over a century, the national pavilion system has been a crucial mechanism sustaining the Venice Biennale's distinctive character while reinforcing national identities. However, since the late 1960s, particularly in the wake of the 1968 student movements, this system has faced criticism, leading to introspective discussions and the gradual incorporation of post-national perspectives. The 1990s marked a period when such shift gained momentum, with transnational practices proliferating through exhibitions and works.

The Korean Pavilion, itself a cultural and institutional product of the 1990s, remains the most recent addition to the Giardini and is often referred to as the "last pavilion." It is only the second pavilion built by an Asian nation after Japan. Conceived as a product of the post-Cold War era, it had to accommodate various social and political implications. The resulting unique form of the Korean Pavilion distinguishes itself from the architectural styles of other national pavilions in the Giardini. Many artists and curators have noted that the Korean Pavilion feels more like a house than a white cube gallery space.[1] This is because the Korean Pavilion needed to fulfill not only its function as a temporary exhibition space but also livability as a public space for citizens. This inherent publicness of the Korean Pavilion resonates with recent tendency in the Biennale, which emphasize inter-pavilion solidarity and the opening up of pavilion spaces. Meanwhile, the recent announcement of a new national pavilion—the Qatar Pavilion—signals upcoming geopolitical changes within the Giardini. This presents a markedly different situation from thirty years ago when even securing land for a new pavilion was challenging. The national pavilions of Giardini now find themselves examining their own temporal existence in light of this shift. The exhibition questions how the Korean Pavilion, soon to shed its title as the last pavilion, can find its role in this transformation.

1. Hyungmin Pai, "Dwelling on the Korean Pavilion," in *Common Pavilions: The National Pavilions in the Giardini of the Venice Biennale in Essays And Photographs*, eds. Diener & Diener Architects with Gabriele Basilico (Scheidegger & Spiess, 2013), 263.

The Significance of the Korean Pavilion's Architecture

> Between paths, between houses, between trees, layering the future upon the past, merging it with the present to create newness.[2]

The Korean Pavilion is the product of collaboration between the late Korean architect Kim Seok Chul and Italian architect Franco Mancuso. It is an atypical steel-framed glass structure that defies conventional exhibition hall designs. The pavilion was created by expanding around a brick building that once served as a public restroom in the Giardini. This "house" was built by resolving nearly impossible site constraints at the time of its construction. The pavilion's characteristics—its transparent body, free-flowing curved plans rather than straight lines, and structure that appears to float lightly above the ground—stand far apart from typical white cube galleries. This distinctive form and layout of the Korean Pavilion stemmed from the stringent requirement that not a single tree on the site could be damaged during construction. Considering Venice is a city built atop foundations made of wooden piles, trees are not merely design constraints but a reflection of the city's survival history. The architecture of the Korean Pavilion, completed with respect for the trees and the land, embodies architectural values of transparency, lightness, and mobility—qualities often associated with the future of architecture itself.

The discovery of trees, which served as the starting point for the *Little Toad, Little Toad* exhibition, paradoxically provided an opportunity to deconstruct and reconstruct the Korean Pavilion. This was made possible not only through oral accounts about the establishment of the Korean Pavilion but also through archival materials that contains detailed records of its design process. In 2023, Franco Mancuso donated the Korean Pavilion design materials to the Korean government. These materials have subsequently stimulated various discussions, including potential pavilion expansion and operational improvements. Additionally in 2024, commemorating the pavilion's 30th anniversary, the Arts Council Korea (ARKO) published *The Last Pavilion*, serving as a research tool for the Korean Pavilion's architectural archive. Furthermore, the ARKO Arts Archive, a national arts archive, is conducting various initiatives to document the pavilion's history, including oral history recordings with Franco Mancuso. While the Mancuso donated archive is being systematically cataloged, we have come to realize that it alone is insufficient for a comprehensive interpretation of the Korean Pavilion. This was partly due to the inability of institutions and researchers to secure materials from Kim Seok Chul, who passed away in 2016. Due to these uncertainties, we chose to explore the Korean Pavilion through the imaginative process of commissioned works.

Exhibition Layout

The exhibition is structured around two main frameworks. The first involves

2. Kim Seok Chul's design sketch memo for the Korean Pavilion. ca. 1990s

caring for the Korean Pavilion from both spatial and temporal perspectives. The goal was to illuminate the spatial origins of the pavilion and reveal the stories embedded within it. Akin to hanging a nameplate on a house or attaching labels to artworks, the curators named the spaces. Captions were provided not only for basic architectural information—such as who designed the pavilion and when it was completed—but also for the surrounding trees that played a foundational role in shaping its form. Meanwhile, documentation compiled from various archives offers a multifaceted perspective on the pavilion's passage of time. At the center of the exhibition, a three-channel video fragments and reassembles moments from different times in the pavilion's history. This narrative features various non-human entities, including the toad that gives the exhibition its title. Their perspectives reflect on not only on the Korean Pavilion but also on fellow national pavilions in the Giardini, prompting visitors to reinterpret the Giardini within a broader context.

The second framework consists of commissioned works by participating artists. Four architects—Kim Hyunjong, Heechan Park, Lee Dammy, and Young Yena—were commissioned to develop new ways of viewing the Korean Pavilion. Known for expanding the boundaries of architecture through diverse media and methodologies including installations and exhibition works, their expertise played a crucial role in determining the spatial arrangement of the works. After conducting site surveys and historical research, the architects deconstructed and reassembled the pavilion. Rather than dwelling on questions of national identity, which has long been a focal point of the Korean Pavilion, their works evoke the shared physical foundations and underlying conditions that occasioned the pavilion's establishment.

At the entrance of the exhibition, Lee Dammy's work summons various natural entities of the Korean Pavilion and informal moments of architectural formation. The works by Young Yena, Heechan Park, and Kim Hyunjong each traverse the pavilion's past, present, and future, engaging with the broader ecological landscape of Giardini—its earth, trees, sky, and sea. These works are installed throughout the pavilion, including its underground space and rooftop, revealing spatial dimensions previously overlooked in the prior exhibitions. Furthermore, they encourage us to continuously observe how the Korean Pavilion meets and connects with its surrounding environment, examining the boundaries between the interior and exterior of the architecture. Rather than proposing physical modifications, the architects position the Korean Pavilion within the ecosystem of national pavilions through methodologies of fiction, simulation, and leaps of interpretation.

From Unbuilding to Living Pavilion

The architects' works, which prompt a reconsideration of the Korean Pavilion's meaning of existence, embody what poet and classicist Anne Carson describes as an "adjectival" attitude—one that allows us to see anew without erasing or

3. Anne Carson, *Autobiography of Red* (Jonathan Cape, 2010), 4.

damaging the original being, akin to creating "latches of being." Carson, in explaining the etymology of adjectives, which originate from meanings of "added" or "appended," notes that "adjectives come from somewhere else."[3] The attitude embodied in the *Little Toad, Little Toad* exhibition is about appending new meanings to reframe our understanding of the Giardini, rather than confining the Korean Pavilion within the premise of being the "last pavilion."

The curators entrusted the four architects with the authority to provide "added" interpretations of the Korean Pavilion. The process of creating the works with the architects was not one of tedious, managerial caretaking but rather a form of exploratory and aspirational care. This is our way of engaging with what we call "matters of attention." As Bruno Latour suggests, organizing such exploratory collective is a way of dealing with non-factual or uncertain issues and connects to the approach of dealing with "things." The Korean Pavilion, as a thing/architecture that has drawn our attention, has come to reveal another mode of its existence through this exhibition.[4]

> Developing a sense of place both enables attention and requires it. That is, if we want to relearn how to care about each other, we will also have to relearn how to care about places. This kind of care stems from the responsible attention that Kimmerer shows us in *Braiding Sweetgrass*, which beyond affecting us by determining what we see, materially affects the very subjects of our gaze.[5]

We, as the fourteenth inheritors of the Korean Pavilion at the Venice Biennale International Architecture Exhibition, hold a responsible attention for this place. As Jenny Odell states, care and attention catalyze forces that lead to substantial change. This force resonates with the purpose of fables that expand imagination by deliberately distorting time, space, logic, scale, and proportion. Looking back at the Korean Pavilion's history led us to consider its relationships with fellow pavilions at the Venice Biennale. As a visual statement manifesting the exhibition's theme, we drew a strikethrough across the COREA signage installed on the pavilion's facade. COREA was the name chosen at the time of establishment, embodying the ideal of a joint exhibition space for both North and South Korea. A strikethrough signifies an act of adding new meaning while preserving existing traces. This gesture reflects our position hoping that the future Korean Pavilion will become a catalyst in dismantling invisible boundaries between national pavilions in the Giardini.

Little Toad, Little Toad explores the potential for transformation of the Korean Pavilion through its own history, and by imagining a detachment from the national identity affixed at its establishment, considers the possibility of a common ground shared with fellow national pavilions. The participating architects' attention shed light on the ground above and beneath the pavilion, as well as its peripheries, calling forth global commons—earth, trees, sky, and sea—that transcend national jurisdiction. These common conditions provide an opportunity to examine what we are connected to. On the Biennale stage,

---

4. For a discussion on commissioning as a curatorial mode of inquiry, see the following article: Trine Friis Sørensen, "A Precarious Construct: The Commission As A Curatorial Mode Of Inquiry," *The Nordic Journal of Aesthetics*, No. 52 (2016), 79-98.

5. Jenny Odell, *How to Do Nothing* (Melville House, 2019), 180.

where artworks and artists typically take the spotlight, we turn our gaze toward the place itself that houses the exhibition. As Simon Sheikh notes, biennales have "the potential for creating a more transnational public sphere," "by being perennial events, both locally placed and part of a circuit."[6] This exhibition aims to contribute to the pavilions' potential to be reborn as places of timeless living rather than as spaces consumed by temporary events. This is also an essential process for pavilions to continue living within the local community amid discussions about the Biennale's crisis. This is our story commemorating the Korean Pavilion's 30 years and the lesson of the "new house" that the toad is passing on to us.

---

6. Simon Sheikh, "Constitutive Effects: The Techniques of the Curator," in *Curating Subjects* (Open Editions/Occasional Table, 2007), 181.

VISUAL CURATING ESSAY

KWAK SEUNG-CHAN

→

1995

A PARADOX BY DESTINY:
NATIONAL PAVILION AS A SPACE FOR ALL

La Biennale di Venezia raised its curtain in 1895. A century later, in 1995, the Korean Pavilion, which would become the last independent national pavilion in the Giardini, opened its doors.

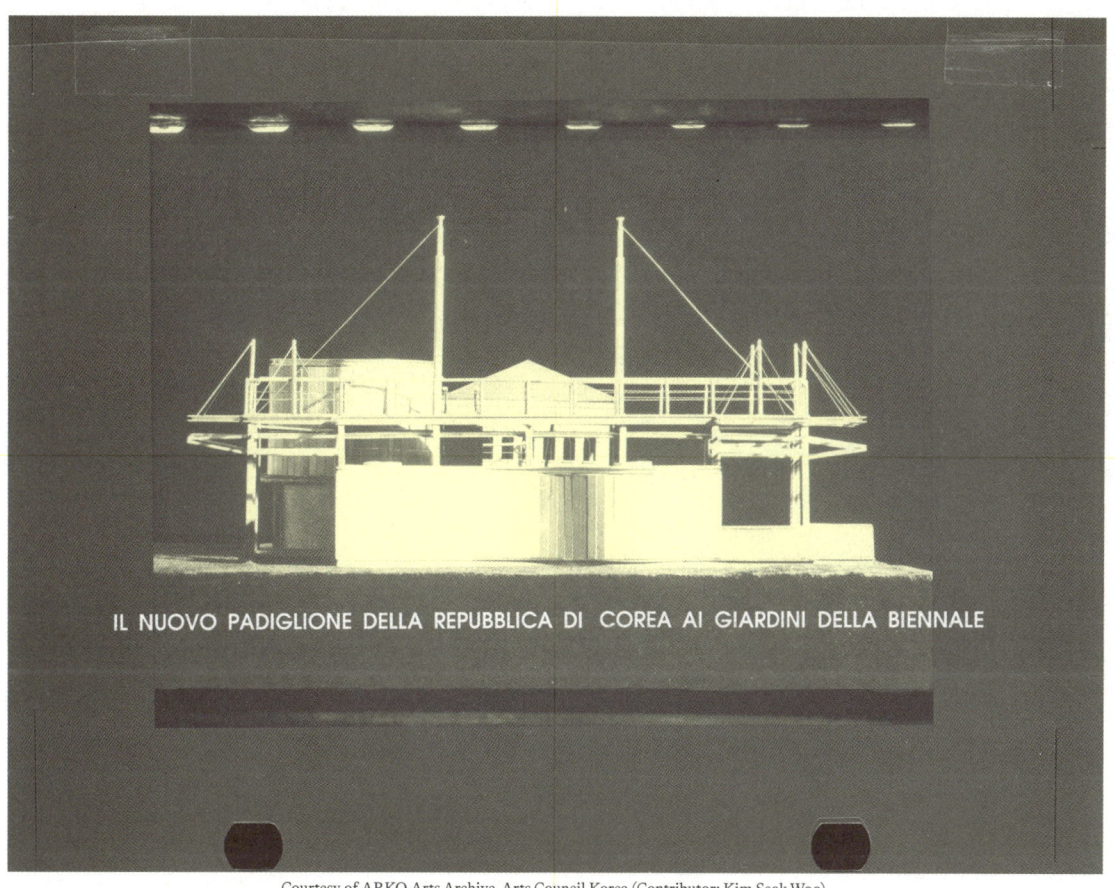

Courtesy of ARKO Arts Archive, Arts Council Korea (Contributor: Kim Seok Woo)

The Korean Pavilion at the time of completion (1995)
©Mancuso e Serena Architetti Associati. Courtesy of ARKO Arts Archive, Arts Council Korea.

The Korean Pavilion during construction (1995)
©Mancuso e Serena Architetti Associati. Courtesy of ARKO Arts Archive, Arts Council Korea.

30 YEARS: As of 2025, the last national pavilion in the Giardini is the Korean Pavilion. Co-designed by Korean architect Kim Seok Chul and Italian architect Franco Mancuso, the Korean Pavilion was completed in 1995 and marks its 30th anniversary this year. In Korea, the age of 30 is regarded as the state-designated age at which architecture dies. This is due to the Urban and Residential Environment Improvement Act, which regulates aging buildings in Korea and permits the reconstruction of structures that have surpassed the 30-year mark. Under this law, Korean houses, particularly multi-unit residential complexes("APT"), are uniformly abandoned the moment they turn thirty, awaiting the fate of demolition and redevelopment. If the Korean Pavilion is considered Korea's 'house,' then the last national pavilion in the Giardini has also reached the age where it must contemplate its mortality.

A NEW ERA FOR THE VILLAGE: In June 2024, La Biennale's President, Pietrangelo Buttafuoco, announced the arrival of a new pavilion in the Giardini — the first in 30 years since the construction of the Korean Pavilion. He stated:

"The Venice Biennale is historically the thermometer of geopolitics. The common home of peoples called to the disciplines of contemporary arts that it is, in its places — and in the succession of a history of 130 years — the Biennale restores the exact measure of an era. The reading of the world and its transformations — where the Pavilions of the many nations present in Venice are the fresco of that precise destiny that is the future — sees in Doha, today, the artistic tension consistent with the reason proper to the Fondazione La Biennale di Venezia."

Here, the network of pavilions is a 'fresco of the future' shaped by global power dynamics, with the emergence of the Qatar Pavilion — the first independent national pavilion from the Middle East in the Giardini — reflecting La Biennale's role as a "thermometer of geopolitics."

As the Korean Pavilion, the last of its kind, approaches a moment of contemplating its mortality, a new pavilion comes to life. If the Giardini is conceived as a 'village' of national houses, its history and system are transitioning into a new chapter.

THE LAST PAVILION AND THE FIRST PAVILION:
The first neighbor invited to this
village was Belgium. Built in 1907 — 88
years before the Korean Pavilion — the
Belgian Pavilion marked the beginning of
national pavilions in the Giardini. At
the time, Belgium had yet to celebrate
a full century of independence, yet
it had rapidly amassed wealth through
early industrialization, becoming one
of Europe's most economically advanced
nations. Belgium was also a notable
late entrant and a rising force in
colonialism: the monarch at the time
was Leopold II, who had established
and privatized the Congo Free State,
perpetrating brutal oppression. Against
the backdrop of a relatively short
history as a nation, rapidly accumulated
wealth, and imperialistic power that
seemed to be accelerating as if to catch
up, a desire for cultural expression of
national identity began to emerge in
Brussels during the late 19th century.
The architect and decorative artist
Léon Sneyers, who designed the Belgian
Pavilion, worked within this cultural
zeitgeist. His methodology was shaped by
his mentor, Paul Hankar, who, along with
contemporaries Victor Horta and Henri
Van de Velde, strived to unearth a new
architectural language that diverged
from the "non-Flemish" neoclassicism.
Though multiple renovations have since
rendered it an entirely different
structure, the first national pavilion in
the Giardini initially aimed to express
national identity through the geometric
Art Nouveau and Symbolist aesthetics of
its time.

The Belgian Pavilion at the time of completion (1907)
Courtesy © Archivio Storico della Biennale di Venezia, ASAC

> Through their external expression, the pavilions competed as representations of empires and nations with styles and scales that demarcated the grounds of the Giardini for decades ahead of the arrival of architecture as a subject. In that sense, and until the 1970s when Vittorio Gregotti brought architecture into different venues in Venice, architecture played its conventional role towards the other arts: as framework.
>
> — Hashim Sarkis, "The Artistic Directors," *The Disquieted Muses: When La Biennale di Venezia Meets History* (2020)

Palazzo dell'Espozione at its construction (1895)
Courtesy © Archivio Storico della Biennale di Venezia, ASAC

Façade of Palazzo dell'Espozione at the time of completion (1895)
Courtesy © Archivio Storico della Biennale di Venezia, ASAC

THE EXPRESSION OF IDENTITY: Indeed, the architectural manifestation of national identity has been inherent to the village since its inception. Of course, Italy, the conservative — or classically anchored — "host," stood apart from other nations in this regard. Consider the oldest existing building in the Giardini, originally constructed for the first Biennale in 1895: the Palazzo dell'Esposizione, once known as Pro Arte. Now referred to as the Central Pavilion, it has long served as Italy's national pavilion as well. True to its institutional role, it was designed by Enrico Trevisanato, the Venetian Council's architect. However, commissioners deemed the initial façade lacking in expressive grandeur. Consequently, artists Bartolomeo Bezzi and Mario De Maria were commissioned to enhance its design. The resulting architectural face of Italy leaned heavily toward a Greek-influenced neoclassicism. Elevated on a high podium, Ionic columns supported an ornate pediment and frieze. At the pediment's apex stood a winged female figure, flanked by the aquila — a clear nod to the Roman legacy.

Façade of Palazzo dell'Espozione after early renovation, with the winged lion of Venice (1907)
Courtesy © Archivio Storico della Biennale di Venezia, ASAC

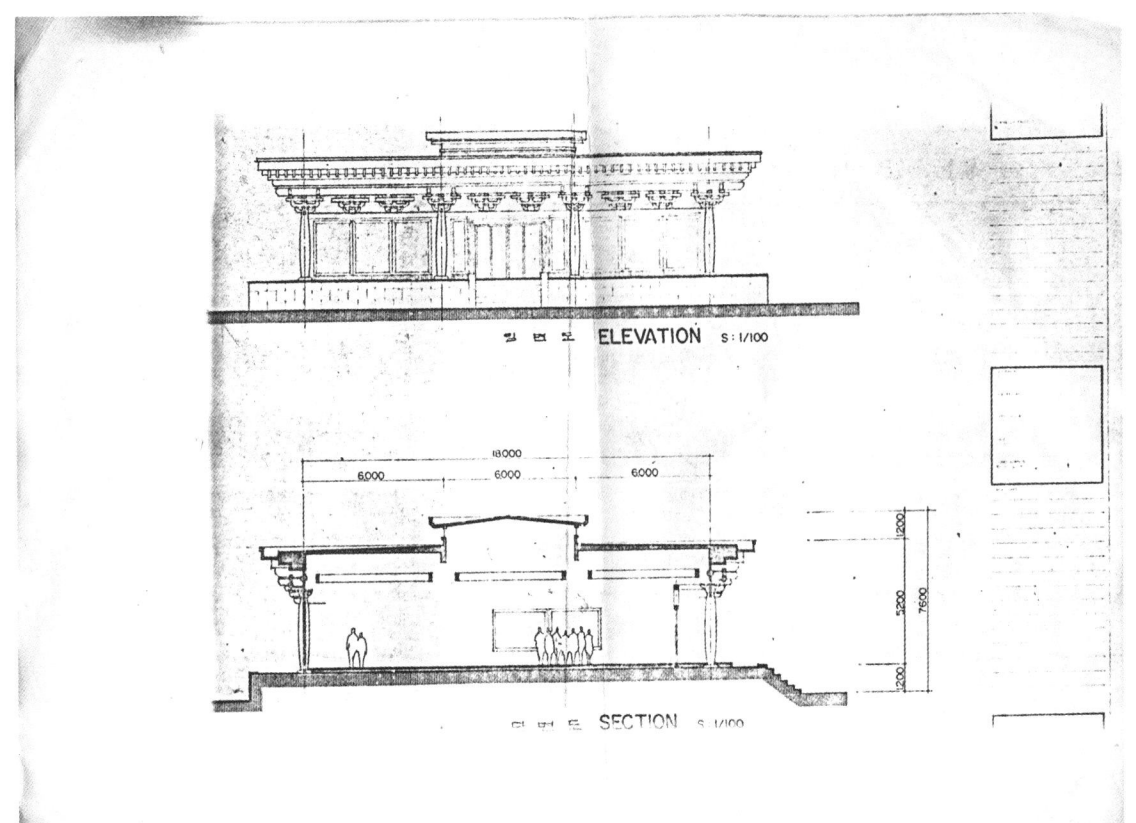

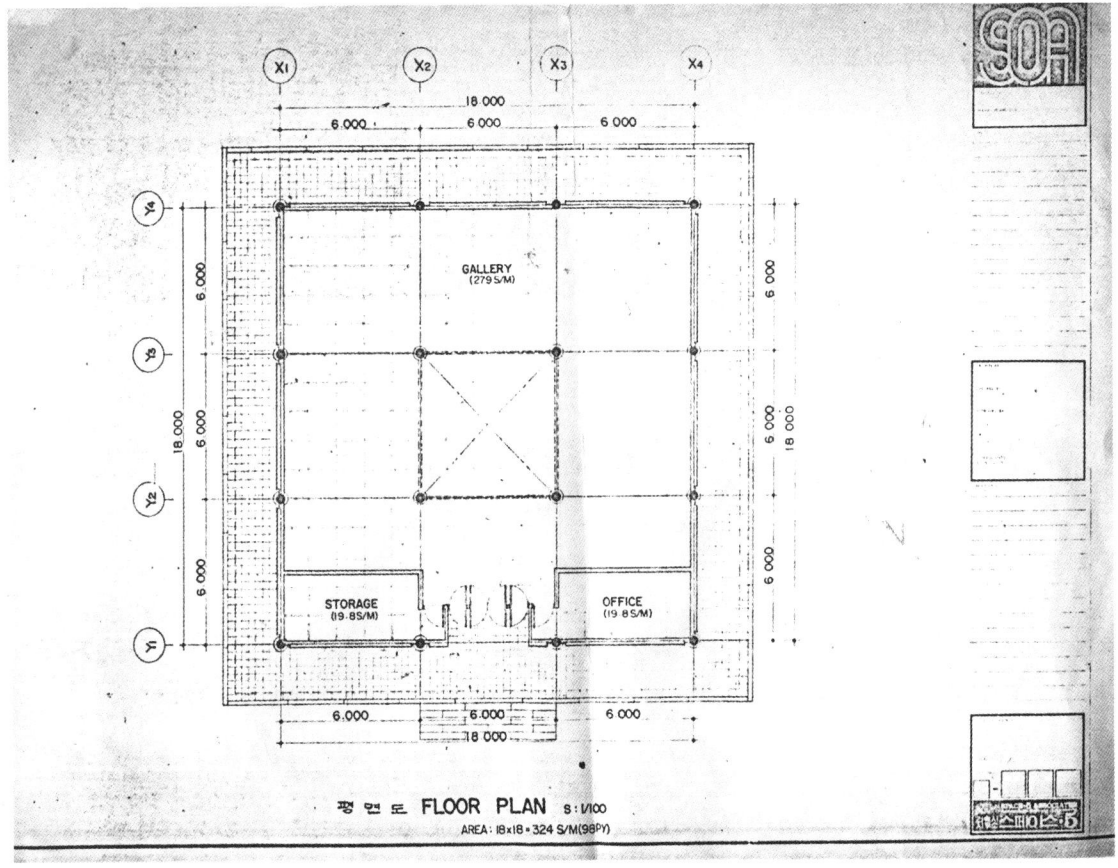

The earliest surviving Korean Pavilion design proposal (initially submitted to Paolo Portoghesi in 1987, photocopy resubmitted to Giovanni Carandente in 1988). The lower right corner of the drawing is marked as designed by "스페이스·5(SPACE·5)." Courtesy © Archivio Storico della Biennale di Venezia, ASAC

Embassy of the Republic of Korea
Rome

Technical Data for the Construction of an Independent Pavilion within the Venice Biennale.

Work by the Korean Fine Arts Association.

Under the patronage of the Ministry of Culture and Information.

- Occupied area: 18 X 18 = 324 m²
- Floors: one (ground)
- Construction material: reinforced concrete
- Facade: stone
- Columns: reinforced concrete
- Steps: stone
- Roof: reinforced concrete
- Walls: blocks and bricks
- External decorations: wood
- Construction period: 6 months
- Cost estimate: 200 million won, equal to approximately 300 million lire. All costs will be covered by the Korean Fine Arts Association.
- Purpose: exhibition of Korean paintings and sculptures.

The bottom image is the renovated German Pavilion by Ernst Haiger
from *Die deutschen Beiträge zur Biennale 1895-2007* (2007)

BIENNALE 1938

Präsident: Giuseppe Volpi
Generalsekretär: Antonio Maraini
Deutscher Kommissar:
Adolf Ziegler

Deutscher Pavillon:

Malerei:
Thomas Baumgartner, Carl Blos, Hans Adolf Bühler, Wilhelm Dachauer, Bernhard Dörries, Franz Eichhorst, Rudolf Hermann Eisenmenger, Leo Frank, Constantin Gerhardinger, Hermann Gradl, Willy ter Hell, Ewald Jorzig, Julius Paul Junghans, Willy Kriegel, Georg Lebrecht, Werner Peiner, Leo Samberger, Hans Schachinger, Georg Siebert, Ferdinand Spiegel, Wilhelm Schmurr, Raffael Schuster-Woldan, Karl Storch d.J., Hermann Tiebert, Karl Walther, Adolf Wissel, Max Zaeper, Adolf Ziegler

Grafik:
Carl Blos, Rudolf Jettmar, Switbert Lobisser, Klaus Richter, Karl Storch d.J., Adolf Ziegler

Plastik:
Arno Breker, Michael Drobil, Alfred Hofmann, Ulfert Janssen, Fritz Klimsch, Ferdinand Liebermann, Josef Müllner, Kurt Schmid-Ehmen, Josef Thorak

THE MODEL HOUSE OF A NATION THAT WAS NEVER ACCEPTED: In 1987, before the construction of the present-day Korean Pavilion, the first surviving design proposal was prepared by the Korean Fine Arts Association and submitted to the President of Biennale, Paolo Portoghesi. This early proposal followed the lineage of national identity-driven architectural expressions exemplified by the first pavilions in the village: the design essentially identified the upper structure of Korean traditional wooden architecture as the defining element of national architectural identity. Like many contemporary Korean architects of the time who were compelled to engage with the question of "Korean-ness," the proposal resorted to a pragmatic solution: emulating the interlocking joinery of "gong-po" (brackets) using reinforced concrete — embracing a modern material while maintaining the aesthetic vocabulary of tradition. In some ways, the design resembled Kim Swoo Geun's 1967 Korean Pavilion for the Osaka Expo, built exactly two decades earlier. However, from the perspective of modernizing Korean-ness, the 1987 proposal leaned more toward direct imitation of tradition, frankly reflecting the persistent aspiration for a "Korean" modern architecture that had dominated architectural discourse for the previous twenty years.

Had it been realized, the pavilion would have found itself in intriguing dialogue with the neoclassical national pavilions of Western powers that surround the current Korean Pavilion — such as the Troost-esque fascist neoclassicism of the German Pavilion, renovated by Ernst Haiger in 1938, or the Renaissance-inspired British Pavilion designed by E.A. Rickards in 1909.

Yet, Korea was a latecomer to the international stage of the arts, and the eyes of the era were changing. Portoghesi reportedly rejected the 1987 proposal on the grounds of its overtly nationalist aesthetic. It is an intriguing reaction, given that Portoghesi himself had curated *La presenza del passato* for the inaugural Venice Biennale of Architecture in 1980. Did he recognize that the emulated wooden structure had emerged in an entirely different context — one in which the Western strategy of playful historical pastiche could not hold the same weight?

A NEW ERA: In 1993, just six years after the first attempt failed, discussions were reignited. This time it was centered around Nam June Paik, Kim Seok Chul, and the Korean Ministry of Culture. By then, the shift in worldview had already been completed: In the interim, the Berlin Wall had fallen in 1989, and the Soviet Union had dissolved in 1991. The post-Cold War condition called for a new optimistic imagination of the "global village." At the same time, Venice Biennale, approaching its centennial in 1995, was contemplating its evolving role befitting the new era. By then, the notion of nationality-architecture, projecting a country's character through physical style, would have seemed like a relic of a bygone era.

> The 1990s brought epoch-making changes that upset the entire global balance of power. This was the decade which followed the fall of the Berlin Wall and the end of the Cold War, the end of the standoff between the great Western and Soviet blocs and the emergence of a new borderless European community. It was also a decade of crisis for the nation state, a founding concept for the Biennale: the exhibition model presented at the Venetian institution was an outgrowth of the nineteenth-century world's fairs that trumpeted the power of colonial empires. Representing national identity through pavilions was another highly symbolic demonstration of the power that nation states attempted to exert in the twentieth century.
>
> — Cecilia Alemani, "90s: From Nation States to a Global Biennale," *The Disquieted Muses: When La Biennale di Venezia Meets History* (2020)

'The Red Pavilion' by Ilya Kabakov, installed in the backyard of the Russian Pavilion in 1993 International Art Exhibition. Courtesy of Ilya and Emilia Kabakov.

THE PREDICAMENT OF NATIONAL PAVILIONS: 1993, the first Biennale of the new era. The Russian Pavilion, a piece of architecture steeped in religious symbolism, was no longer representing the Soviet Union. In its backyard a small "new pavilion" was erected, and this modest structure sharply exposed the predicament that the nationality-pavilion model was facing. The structure was 'The Red Pavilion' by Ilya Kabakov, installed as part of the Russian Pavilion exhibition *Community of Independent States*. Visitors entering the Russian Pavilion were guided through an interior strewn with discarded debris and remnants of the past before emerging into the backyard. There, they finally encountered a crude wooden shack adorned with red stars and flags, while the faint strains of 'The Internationale' streamed out from a worn-out gramophone.

From beyond the fence, a winged lion silently observed the scene. Once perched atop the Accademia, symbolizing the imperial grandeur of Venice, it had since been displaced. Now, it had come to rest at the entrance of a narrow path that would one day lead to the Korean Pavilion.

'Marco Polo' by Nam June Paik, exhibited in the Giardini for the German Pavilion of the 1993 Biennale. ©Mancuso e Serena Architetti Associati. Courtesy of ARKO Arts Archive, Arts Council Korea.

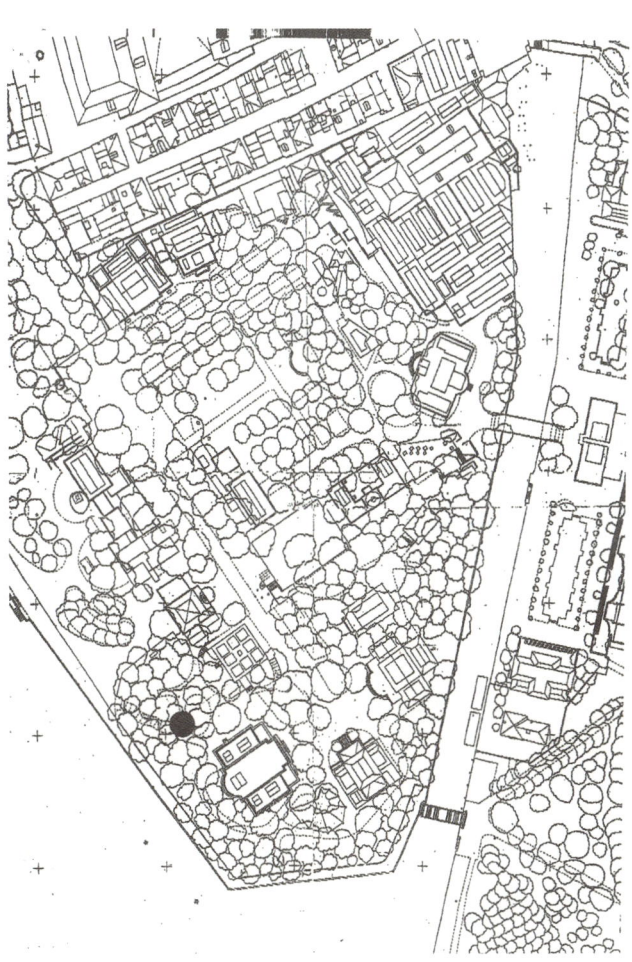

The site of the Korean Pavilion, marked in a black dot. Nam June Paik's 1993 German Pavilion works were scattered on the site. ©Mancuso e Serena Architetti Associati. Courtesy of ARKO Arts Archive, Arts Council Korea.

NAM JUNE PAIK AS THE POINT OF ORIGIN: Meanwhile, in the neighboring German Pavilion, a similar scene of ruin unfolded: Hans Haacke shattered the pavilion's floor, forcing visitors to walk across the "ruins of the country." However, unlike the Russian Pavilion, the German Pavilion did not merely offer contemplation or a bitter smile at the end of the grand confrontation between the First and Second Worlds. This was thanks to Nam June Paik, who formed a pair with Haacke. If Haacke was an artist who was "born in Germany and moved westward (to America)," Paik was one who was "born in the East and moved westward (to Germany)." Paik literally "stepped outside" of the ruined German Pavilion — onto the very wooded hill where the Korean Pavilion would be built two years later. There he scattered historical figures who had traversed borders, reincarnating them as his signature video sculptures. These robotic figures — drawn from completely different eras, holding vastly different statuses, and even varying in their historical authenticity — were loosely connected beyond the trees. In doing so, the future site of the Korean Pavilion was reimagined as a speculative archaeological-futuristic space, a playful counter-world of the grand narrative. The inhabitation of TV-Marco Polo, TV-Alexander the Great, TV-Genghis Khan, and TV-Tatars reterritorialized the forest into a world of crossings, powered by the advancement of long-distance communication technologies.

News article reporting Nam June Paik's Golden Lion Award of 1993 International Art Exhibition. 'Tangun, Scythian King' on the right. 'Alexander the Great' on the left. Courtesy © *The Chosun Daily*, June 15th, 1993. ("Nam June Paik awarded the Golden Lion for Best National Pavilion at the Venice Biennale.")

And there, among them, stood Tangun, staring at the Adriatic Sea. Tangun, the mythical progenitor of the Korean people, is a constructed nationalistic symbol — yet Nam June Paik's approach was, as always, unconventional. Here, Tangun was reimagined as a nomadic ruler of the Scythians, a Mongolian steppe people. This positioned Tangun as a transcultural figure who had traversed East and West, spending time in Greece before journeying to ancient Gojoseon. In the free port of Venice, the forefather of the Korean people is suddenly inverted: He was no longer a static national icon but a continental traveler gazing back at the boundaries he had once crossed.

With his work for the 1993 German Pavilion, Nam June Paik won the Golden Lion for Best National Pavilion.

Canaletto, 'The Return of the Bucintoro to the Molo on Ascension Day' (1730), depicting Sposalizio del Mare (The Marriage with the Sea) of Venetians.

THE LIQUID CITY: Though the work for the German Pavilion marked a defining moment, the liquefaction of boundaries had long been a central theme in Paik's artistic practice. In 1975 — while in Korea, the Park Chung Hee dictatorship was busy inventing new symbols of Korean-ness — Paik participated in *Proposte per il Mulino Stucky*, an exhibition curated by Vittorio Gregotti. There, he presented 'How to get free oil,' a proposal to transform the Venetian island of Giudecca into a free port of information, a place where multiple cultures would converge and intermingle. His vision was to extend Venice's historical role as a hub of exchange into the modern era. The intermingling of different powers and cultures — the essence of a cosmopolitan city and a liquid space — is, indeed, one of Venice's fundamental urban characteristics. Francesco Petrarch once described Venice as a "mundus alter" (another world), a phrase that encapsulated its liminal position: neither fully Western nor fully Eastern. From the perspective of medieval Christian Europe, Venice appeared almost more aligned with the Islamic and Byzantine worlds than with the Latin West. As a maritime mercantile empire, Venice flourished precisely because of its refusal to take rigid ideological positions. Its prosperity stemmed from pragmatism, from its ability to navigate and negotiate between cultures. In this sense, Venice is not only 'a city of water' but, metaphorically, a city that is itself water — a place in constant flux, shifting forms, and flowing between worlds. La Biennale di Venezia stands with one foot in the realm of nationalism and the other in the fluid cosmopolitanism.

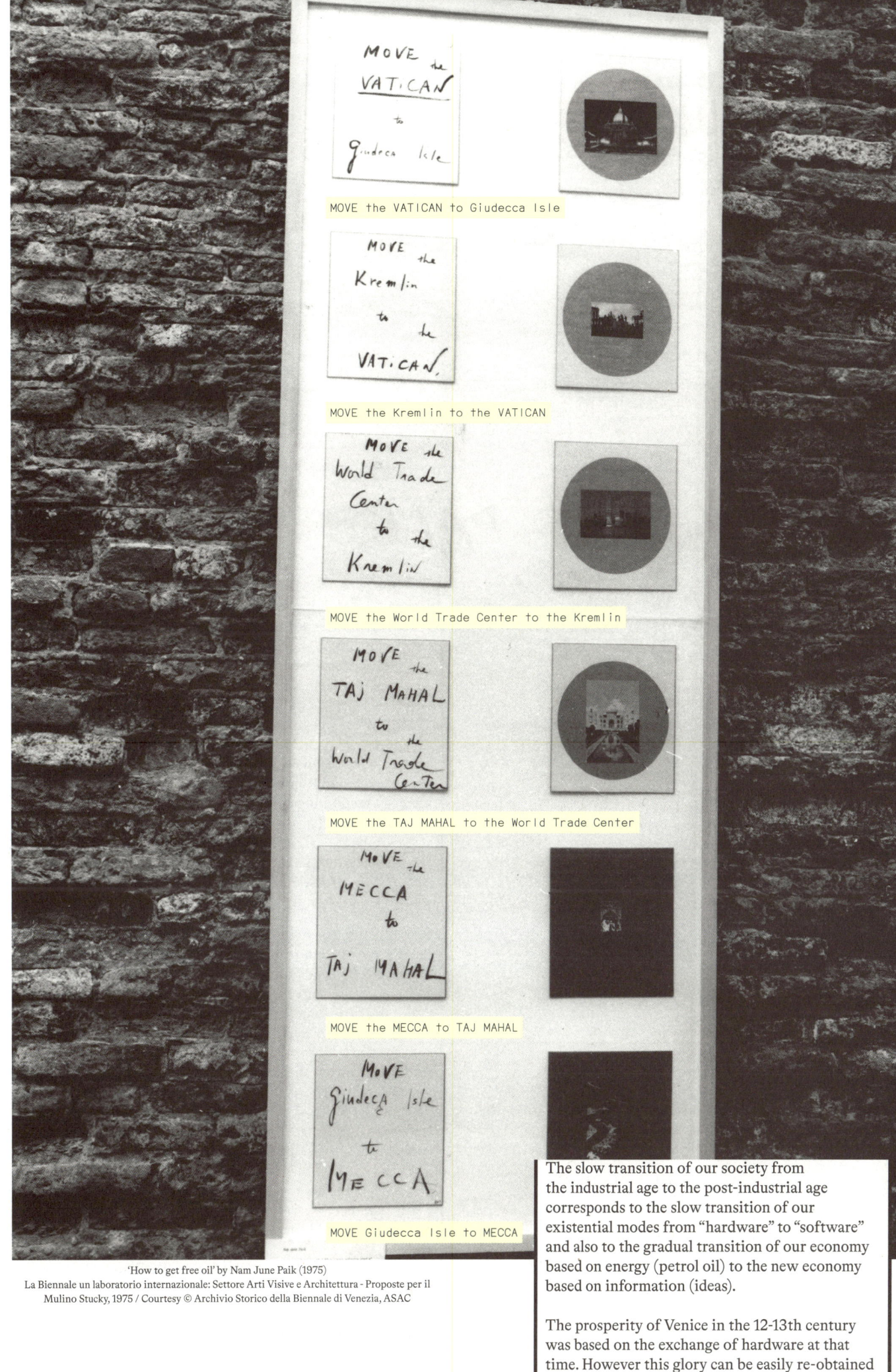

MOVE the VATICAN to Giudecca Isle

MOVE the Kremlin to the VATICAN

MOVE the World Trade Center to the Kremlin

MOVE the TAJ MAHAL to the World Trade Center

MOVE the MECCA to TAJ MAHAL

MOVE Giudecca Isle to MECCA

'How to get free oil' by Nam June Paik (1975)
La Biennale un laboratorio internazionale: Settore Arti Visive e Architettura - Proposte per il Mulino Stucky, 1975 / Courtesy © Archivio Storico della Biennale di Venezia, ASAC

The slow transition of our society from the industrial age to the post-industrial age corresponds to the slow transition of our existential modes from "hardware" to "software" and also to the gradual transition of our economy based on energy (petrol oil) to the new economy based on information (ideas).

The prosperity of Venice in the 12-13th century was based on the exchange of hardware at that time. However this glory can be easily re-obtained by the exchange of software at the isle of Giudecca.

— Nam June Paik, "Idea for Giudecca Isle," *A Propos of Mulino Stucky* (1975)

A fax letter sent by Kim Seok Chul to Franco Mancuso on June 14th, 1994.
©Mancuso e Serena Architetti Associati. Courtesy of ARKO Arts Archive, Arts Council Korea.

THE LAST PAVILION FOR THE LAST DIVIDED NATION: Fast forward twenty years. In the mid-1990s, as the seemingly eternal ideological structures of the world order collapsed and the rise of information technology promised to dissolve all physical boundaries, one border still remained unbroken. It was a division between North and South Korea. Following China's economic reforms and the reunification of Germany, the Korean Peninsula stood as the last frontline of the old system. With the Soviet Union also moving toward reconciliation with the capitalist world, North Korea, feeling increasingly vulnerable, accelerated its nuclear development. In 1993, it announced its withdrawal from the Nuclear Non-Proliferation Treaty (NPT), and by June 1994, the peninsula teetered on the brink of war. The North Korean nuclear crisis is now a normalized status quo. However at the time it was an urgent global concern, appearing as one of the final unresolved threats left by the Cold War. This contemporary context served as a decisive geopolitical condition that made the establishment of the Korean Pavilion a reality, amid competition from numerous other candidates for an independent national pavilion.

THE UTOPIA-DREAMING PAVILION TO IGNITE WORLD PEACE: The key turning point that translated the geopolitical conditions into the establishment of the Korean Pavilion was the idea of a joint North and South Korean Pavilion. This optimistic vision proposed the creation of a single national pavilion for the divided Korean Peninsula, serving as a space where the two nations could engage in cultural and artistic exchange while fostering hopes for future reunification. This was also a proposal that Venice, historically a crossroads where the world converged, could play a modern role as a space for integration, contributing to the resolution of the last ideological division of the 20th century. The concept captivated Achille Bonito Oliva, the director of the Visual Arts sector at the 1993 Biennale, who became a strong advocate for the project. This idea ultimately seemed to persuade Mayor Massimo Cacciari, key officials of the city of Venice, as well as the Italian government. As both architect Kim Seok Chul and art critic Yongwoo Lee, who closely assisted in the pavilion's establishment, have both testified, the person who first conceived the idea of a joint pavilion was none other than the visionary Nam June Paik. Considering that twenty years earlier, he had proposed "moving the Kremlin to the Vatican and the World Trade Center to the Kremlin," this idea seemed comparatively less radical.

After two months with the mayor's seat left vacant, Massimo Cacciari finally took office. Fortunately, he was a close friend of Professor Rinio Bruttomesso. Nam June Paik sent the newly appointed mayor a pictorial letter, stating:

"This is your opportunity to win the Nobel Peace Prize. Imagine how historically significant it would be if, at the site of the Venice Biennale, soon to mark its 100th anniversary, North and South Korea, the last nation still divided by ideology, could come together and address the nuclear issue through culture."

— Kim Seok Chul, "Venice Biennale Korean Pavilion," *Architecture for All, City for All* (2013)

Permission to construct the Korean pavilion occurred in the context of a politically idealistic desire for unification between North and South Korea. That is, the pavilion's construction conjured up the beautiful idea that North and South Korea, using the pavilion together, could resolve their political tension through art. ⋯ However, this idea — suggested by the artist Nam June Paik in order to argue for the legitimacy of the pavilion's construction — startled me, the architect Seok Chul Kim, and others who had been promoting the establishment of the Korean Pavilion. ⋯ The Italian Government agreed to the possibility that Venice could become the stage for the realization of such unification. ⋯ Even before work on the Korean Pavilion began, Paik emphasized the necessity of its construction by going so far as to say that there was a chance for the government of Venice to receive the Nobel Peace Prize.

— Yongwoo Lee, "Heroes of the Korean Pavilion at the Venice Biennale," *Kimsooja, To the Breath: Bottari* (2013)

Il Ministro degli Affari Esteri

LA BIENNALE DI VENEZIA
19 LUG. 1994
PROT. N. 5640 P
ARRIVO

Pres. ASI Uff. Tecn
19 LUG. 1994

113/ 2523
Roma, 7 LUG. 1994

Gentile Presidente,

Le scrivo in relazione al progetto di costruzione, negli spazi della Biennale, del padiglione espositivo dedicato alla Corea, la cui inaugurazione potrebbe avvenire in occasione della ricorrenza del Centenario dell'Esposizione prevista per il prossimo anno.

Desidero assicurarLe fin d'ora che l'iniziativa - che sta particolarmente a cuore al Ministero della Cultura coreano ed a tutto il gruppo dei promotori - gode dell'appoggio di questo Ministero, soprattutto in considerazione dell'impatto sull'opinione pubblica in un momento così delicato nelle relazioni tra le due Coree.

Credo fermamente che operazioni in campo culturale e artistico - ove sono meno sentiti i conflitti etnico-politici - possano essere validi strumenti per superare le tensioni.

So che i promotori hanno già sottoposto alla Sua attenzione il progetto, per il quale, mi assicurano, è stato già predisposto il relativo finanziamento.

Confido nel Suo aiuto al fine di accelerare l'iter per la concessione dell'autorizzazione necessaria all'avvio definitivo dei lavori.

Mi è gradita l'occasione, gentile Presidente, per inviarLe i miei più cordiali saluti.

Antonio Martino

Prof. Gian Luigi RONDI
Presidente
Biennale di Venezia
San Marco - Ca' Giustinian
30124 VENEZIA

> I am writing to you regarding the project for the construction, within the spaces of the Biennale, of the exhibition pavilion dedicated to Korea, whose inauguration could take place on the occasion of the centenary of the Exhibition scheduled for next year.
>
> I wish to assure you from now on that this initiative — which is particularly dear to the Korean Ministry of Culture and to all the group of promoters — enjoys the support of this Ministry, especially considering its impact on public opinion at such a delicate moment in relations between the two Koreas [North and South Korea]. I firmly believe that operations in the cultural and artistic fields — where ethnic-political conflicts are less felt — can be valid tools for overcoming tensions.

A letter from Antonio Martino (Italian Minister of Foreign Affairs) to Gian Luigi Rondi (President of the Biennale), July 19th, 1994.
ASAC, Fondo storico, Progetti speciali b. 6120
Courtesy © Archivio Storico della Biennale di Venezia, ASAC

The letter sent directly by Antonio Martino, then Italy's Minister of Foreign Affairs, to Gian Luigi Rondi, then President of the Biennale, serves as clear evidence of the impact and significance of the joint North and South Korean Pavilion idea.

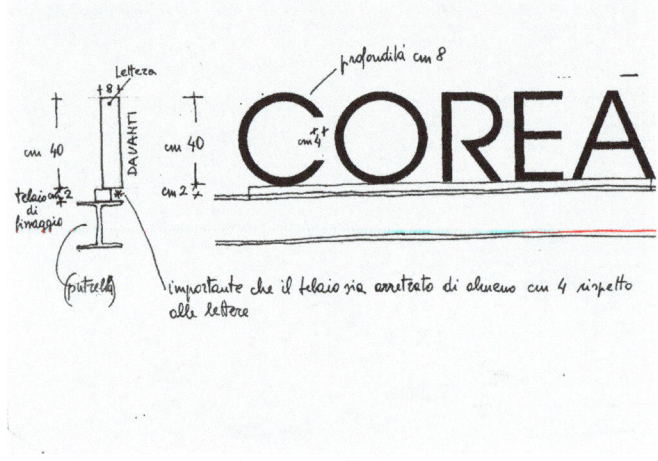

Design for the signage of the Korean Pavilion
©Mancuso e Serena Architetti Associati. Courtesy of ARKO Arts Archive, Arts Council Korea.

Because it was not built as a sole pavilion for South Korea, a sign reading "COREA" was placed above the entrance of the pavilion. This was a deliberate choice, signaling the intention for the pavilion to encompass not just "Corea del Sud" but also "Corea del Nord."

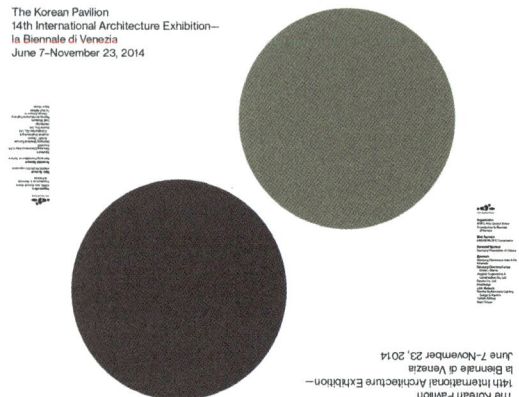

The poster of *Crow's Eye View: The Korean Peninsula*, the Korean Pavilion for the 2014 Venice Biennale of Architecture, designed by Sulki & Min.

THE PROMISED APEX: At the 2014 Venice Biennale of Architecture, the Korean Pavilion attempted its first-ever joint exhibition between North and South Korea. Although North Korea's participation was never realized, the idea itself culminated in the pavilion's first ever Golden Lion award. It was only fitting that the long-forgotten vision that had originally brought the Korean Pavilion into existence returned as a defining climax in its exhibition history.

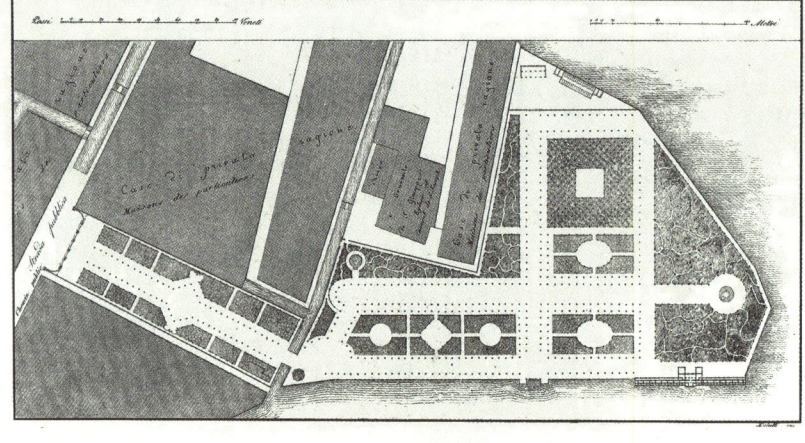

Giannantonio Selva, Plan of Giardini Pubblici (1807)

GIARDINI, NOT PUBBLICI: Besides the global shifts following the collapse of the communist bloc, another powerful context existed on a smaller yet equally significant scale: the long-standing question of publicness in Giardini. If the political conditions of the post-Cold War era explain why the Korean Pavilion was able to be built, then the issue of Giardini's publicness, as a local condition, explains why it was built in this particular form.

The park now known as Giardini di Castello was originally named Giardini Pubblici, or "Public Park," when it was established by decree of Napoleon Bonaparte in 1807. Designed from a modern, planar perspective, the park was conceived as a spatial device to enlighten the medieval city and was thus open to all citizens. Today, however, no one refers to the Giardini with the term "Pubblici" anymore. More than half of the park's area has become isolated from the city's urban fabric, effectively turning it into an enclave. The enclavization of the Giardini began with the introduction of the national pavilion system. In 1907, when the first independent national pavilion — the Belgian Pavilion — was built, fences were erected around it to restrict entry to ticket holders only. The Hungarian Pavilion and the Bavarian Pavilion (now the German Pavilion), which followed shortly after, adopted the same approach. While the national pavilion system became one of Venice Biennale's key sources of revenue in its early years, its institutionalization gradually stripped Venetian citizens of their only large-scale public green space.

In 2001, the Council of Venice approved a detailed urban plan for the Giardini della Biennale. The plan states: "The original function of the Giardini changed over time as the cultural activities of the Biennale put down roots, removing much of the area from public access and generating certain contradictions with the actual use of the space. … The area of the Biennale is completely encircled by enclosures of various kinds, made of different materials: high brick walls alternate with buildings that stand around the perimeter, whereas a long iron fence delimits the Fondamenta." The urban plan commanded "greater accessibility, in terms of duration and overall use of spaces open to the public," regulating the creation of "new publicly accessible waterfronts along the Rio del Giardini" and "the removal of existing walls and fences in favor of temporary structures used only for the duration of public events."

The Biennale did nothing to demonstrate compliance with this directive, and the City did nothing to enforce it. On the contrary, the Biennale recently doubled its fences towards the Fondamenta, removing the weak points known by the locals as "secret entrances" to the now privatised park.

— Unfolding Pavilion 2023, "Giardini Pubblici, A Quick History"

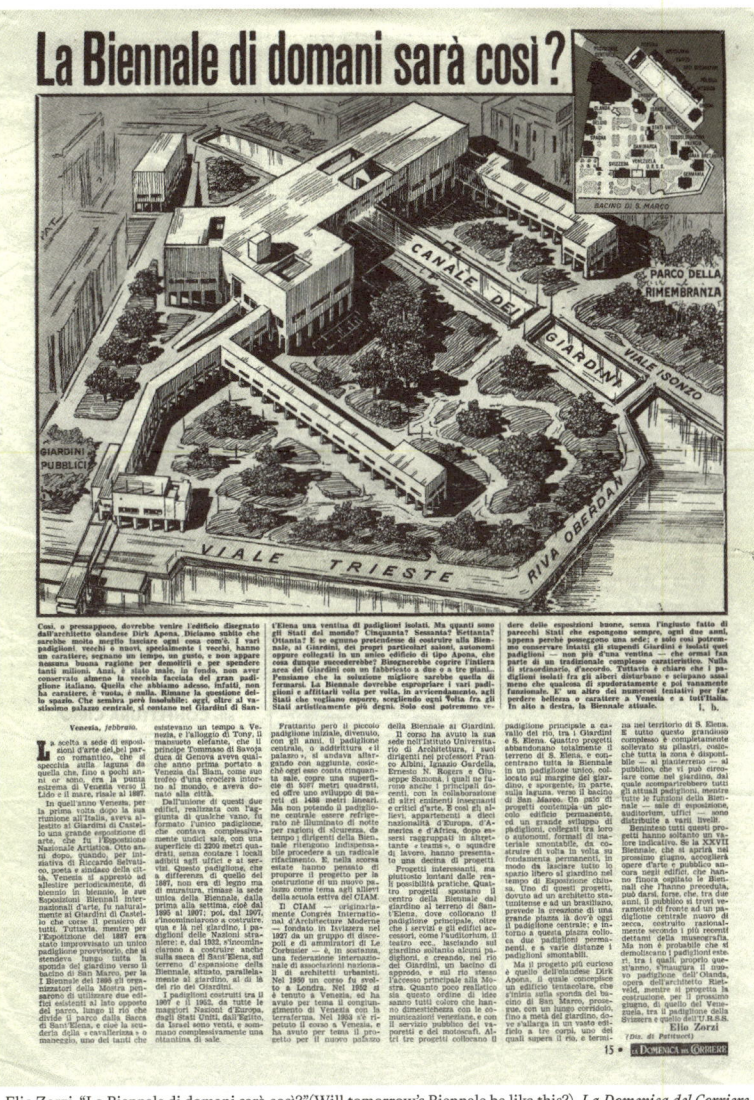

Elio Zorzi, "La Biennale di domani sarà così?"(Will tomorrow's Biennale be like this?), *La Domenica del Corriere*, February 1954. ASAC, Fondo storico, Lavori alle sedi, Padiglioni b. 9, Ritagli stampa 1954-1955
Courtesy © Archivio Storico della Biennale di Venezia, ASAC

Thus, or thereabouts, should come the building designed by Dutch architect Dirk Apona. … The question of space remains. Which seems unsolvable: today, in addition to the vast central palace, there are about twenty isolated pavilions in the Gardens of Sant'Elena. But how many states are there in the world? Fifty? Sixty? Seventy? Eighty? And if each one claimed to build their own particular halls at the Biennale, in the Gardens, autonomous or connected in a single Apona-type building, what then would happen? We would need to cover the entire area of the Gardens with a two or three-story building …

A LONG-STANDING CONCERN OVER PUBLICNESS: The national pavilions in the Giardini expanded without a unified master plan, leading to an uncontrolled sprawl of identity-driven architecture from different nations. As a result, concerns arose early on about the disorderly proliferation of these structures, which, outside of exhibition periods, transformed the park into a desert island of locked and vacant buildings. In 1953, participants in the CIAM (Congrès Internationaux d'Architecture Moderne) Summer School were given the task of recognizing this problem and reimagining the Giardini. Among the various proposals, the one by Dutch architect Dirk Apona drew particular attention. Apona proposed abolishing the individual national pavilion system in favor of a single, unified exhibition hall. In his vision, all existing national pavilions would disappear, and Venice Biennale's entire program — exhibition halls, auditoriums, and offices — would be consolidated into a multi-story structure. This building would be designed as an elevated structure, supported by columns, leaving the ground level entirely open as a public space, allowing citizens to move freely through the Giardini as if it were a garden.

VISUAL - CURATING - ESSAY

Giardini, Steve McQueen, British Pavilion, 2009, 35 mm film, shown as video, high definition, 2 projections, colour and sound (surround), 30' 8"
© British Council Collection, Courtesy Thomas Dane Gallery, London

More than seventy years after Dirk Apona's radical proposal, the issue of the individual national pavilion system and the publicness of Giardini remains unresolved. Steve McQueen's 2009 work 'Giardini,' presented at the British Pavilion, captures the sense of emptiness that overtakes Giardini during the months when the Biennale is not in session. In a landscape void of human presence, resembling a scene from a post-apocalyptic film, nature's entropy reclaims its role as the protagonist.

"Quale Biennale dopo 100 anni? Identità - Prospettive - Riforma"
Courtesy © Archivio Storico della Biennale di Venezia, ASAC

...Fiorella Minervino, wrote in the Corriere della Sera that the Biennale was like a "bedraggled and crushed lady, with such big holes in her veil that her wizened face was no longer hidden" and then reflectively enquired, "then what of the future of the old lady?" The answer was: "one only: close her for 1995. Close the Giardini gates and Palazzo del Cinema doors with courage, together with the appetites of the specialists and merchants."

— Enzo Di Martino, *The History of the Venice Biennale 1895-2005* (2005)

THE HUNDRED-YEAR-OLD LADY: As the Biennale, first opened in 1895, approached its centenary in 1995, it was keenly aware of the milestone and actively preparing for it. Discussions on redefining the contemporary role and function of the Biennale in the new century had already begun as early as 1992. Forums were held in Rome and Venice to gather insights from historians, critics, and former sector directors. In January and February of 1994, marking the 99th anniversary, newly appointed president Gian Luigi Rondi organized an extensive six-day study titled "Quale Biennale dopo 100 anni? Identità - Prospettive - Riforma" (Which Biennale after 100 years? Identity - Perspectives - Reform). The discussions intensified, but no clear consensus was reached. Some took a radical stance, likening the Biennale to a "grand old lady" weighed down by its past, arguing that the only way to properly commemorate its centenary would be to shut it down altogether.

1995: THE YEAR TO OPEN THE BIENNALE: Less than two weeks after "Quale Biennale dopo 100 anni?" concluded, on March 11, 1994, Gian Luigi Rondi made a historic and shocking announcement. For the first time in the Biennale's history, foreigners were appointed as directors of its individual sectors. And not just one, but three. French art critic Jean Clair was placed in charge of the Visual Arts sector, Austrian architect Hans Hollein was appointed to Architecture, and Spanish director Luis Pasqual took over the Theater section. Reflecting on this pivotal moment, art critic Enzo Di Martino described it as "the arrival of the foreigners." Venice Biennale was now opening its doors to the world. What had once been an Italian-hosted gathering of neighboring European powers was evolving into what we now recognize as the international platform for cultural exchange. And that same year, Gian Luigi Rondi and the Biennale authorities selected Korea as the new centennial national pavilion, surpassing other contenders such as Argentina and Portugal, which had maintained longer-standing ties with Italy or had been petitioning for a pavilion for a longer period.

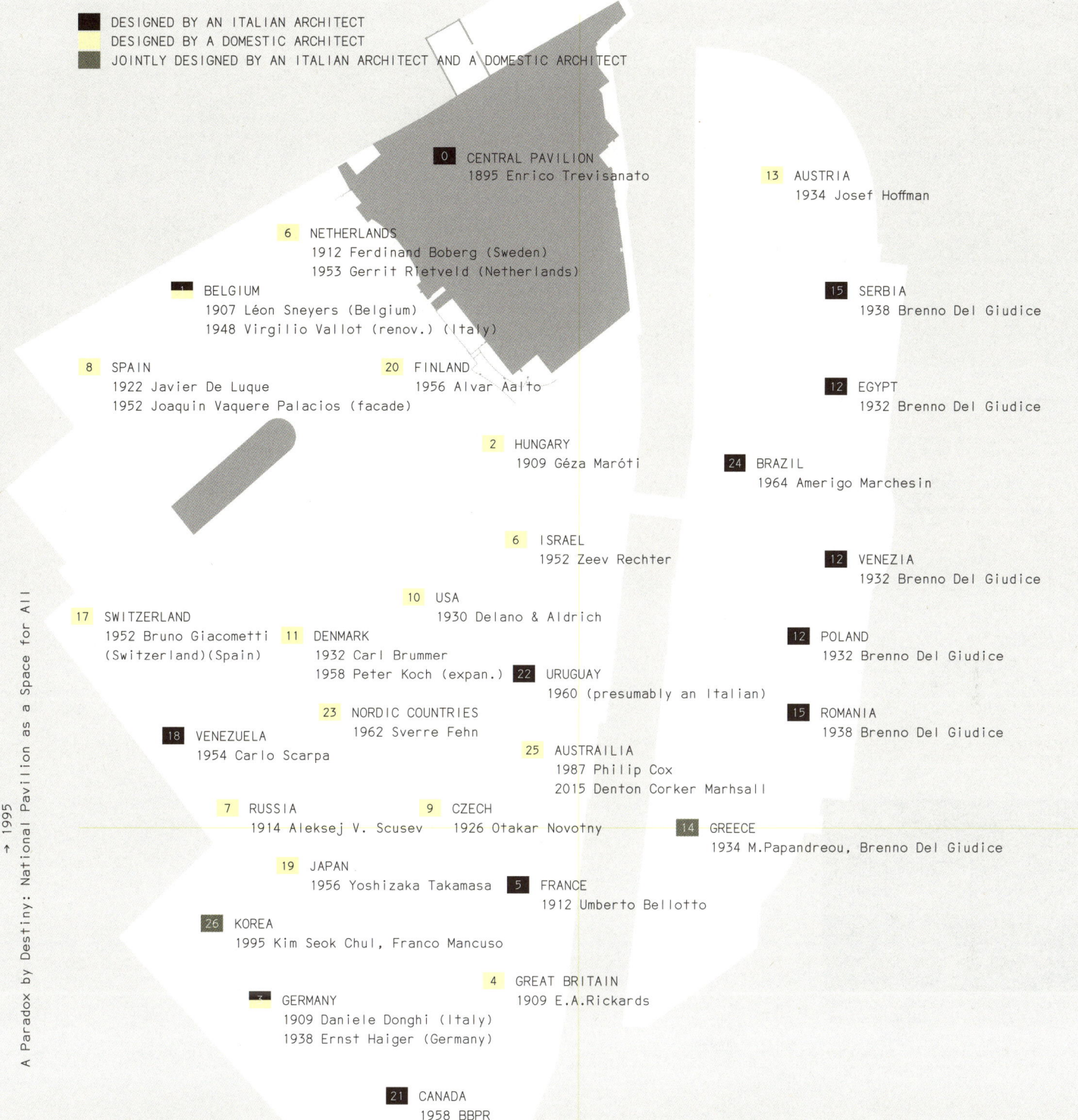

A NATIONAL PAVILION AS AN INTER-NATIONAL JOINT DESIGN: Reflecting the new zeitgeist of a dissolving world order and the corresponding changes within Venice Biennale, the Korean Pavilion was conceived from the outset as an inter-national collaboration between architects from Korea and Italy. This approach was proposed to the Biennale authorities as part of the pavilion's characteristics. (Leveraging the IUAV network through Franco Mancuso also served as a pragmatic strategy for realizing the pavilion.) The co-design between a national architect and a Italian local architect was the first of its kind among all national pavilions in Giardini, with the exception of the Greek Pavilion in 1934. Notably, Kim Seok Chul and Franco Mancuso's collaboration was practical and organic, spanning from the initial concept to completion. This approach was hinting a departure from the old tradition inherited from World's Fairs, in which each pavilions served as exhibits of imperial powers displaying their national identity. The first national pavilion of a new era would follow a different path — one that moved beyond the unilateral physical representation of a single nation.

ARCHIBAN LTD.
───────────────────────────────
1-132 Dong Soong-Dong Jong Ro-Gu Seoul, Korea
Phone 027643072   Fax 027654651

July 8, 1993

By Fax (one page including this one)

Istituto Universitario Di
Architecttura Di Venezia
To : Prof. Franco Mancuso
Fax : 041/5238121

Dear Franco,

I hope you had a nice trip to Japan.
We would like to propose a Korean Pavilion in the Vienale Internazionale di Arte, within the Giardini Pubblici. If that is not possible, somewhere close to it.
We have the support of the Cultural Department of Korea and a sponsor. Since the Korean Government is directly submitting a proposal for this, we need some information and materials.
This would be the beginning of a Joint Venture between you and me in Seoul and in Venice.

Sincerely,

김석철
Kim Seok Chul

A fax letter containing Kim Seok Chul's initial proposal for joint design, written on July 8th, 1993.
©Mancuso e Serena Architetti Associati. Courtesy of ARKO Arts Archive, Arts Council Korea.

FRANCO MANCUSO ARCHITETTO   30135 VENEZIA SANTA CROCE 29 CAMPIELLO MOSCA TEL./FAX 041/5238121

Venezia, 10.09.1993

Dear
Seok Chul
ARCHIBAND LTD
1-132 Dong Soong Korea

fax 0082.2.765.4651

Dear Seok Chul,

thank you very much for your fax. I apologize to you for writing so late, but I've been away from Venice for some weeks, because of a trip in the ex Jugoslavia.

Your news about the work for the Korean Pavillon are very welcome; especially the idea to work as one of the two architects for the project. As you said, it will be a nice, very nice joint venture between the two countries, you and I.

I've visited several times the chosed site, behind the Japanese Pavillon. The place is very nice, because of the presence of a small hill, and many trees. The only problem seems to me that of the existing restrooms: a small building, used as pubblic toilet, but placed exactly on the top of the hill, at the center of the area.

It's necessary to have soon informations about the possibility to remove the restrooms, because it seems very difficult to have a pavillon coexisting with a toilet building.

In any case, the idea to have an underground pavillon is very stimulating. The solution is easy, because of the hill heinght; but the decision about the position of the entrance (and the level) becames very important: a very exciting desing problem.

I discussed the idea of Korean Pavillon with professor Bruttomesso and professor Folin: we get their strong support. Next monday I will met professor Berlanda.

What seems very difficult for me is to give you the cost estimate for the building: it could be necessary to precise some general elements of the project (size, levels, entrance, light, ...); besides, it's necessary to decide something for the restrooms building, and the possibility to move with the project among the many existing trees (protected by the Municipality of Venice).

It could be very nice, and uslful, to have a meeting. Please, tell me if you think to come to Venice, and in any case if you need more informations (on the place, trees, restrooms, etc.).
In the meanwhile, have my best regards. And thahk you again, for all the idea.

Sincerely

(Franco Mancuso)

A fax letter in which Franco Mancuso first expressed his acceptance of the joint design proposal, written on September 10th, 1993. ©Mancuso e Serena Architetti Associati. Courtesy of ARKO Arts Archive, Arts Council Korea.

Kim Seok Chul and Franco Mancuso presenting their design proposal to the Korean Minister of Culture, Lee Min Sup (1994)
©Mancuso e Serena Architetti Associati. Courtesy of ARKO Arts Archive, Arts Council Korea.

MINISTER OF CULTURE AND SPORTS
SEOUL, KOREA

31, August 1994

Dear Prof. Mancuso,

I am in full appreciation on your efforts in the process of realization of Korean Pavilion. I am holding you as a co-architect in high esteem.

I believe that your efforts and achievments contribute to enhance the friendly relationship between Korea and Italy. The Korean Pavilion, as a result of co-work of you and Mr. Kim, will be a symbol of amity of two countries.

The day of laying the foundation stone of Korean Pavilion will be a day of opening a new horizon in the relationship of Italy and Korea. I hope to share the joy with you. You will be one of the most distinguished persons in this event.

Sincerely Yours,

Min Sup, Lee
Min Sup Lee
Minister of Culture and Sports
The Republic of Korea

A letter sent by Minister of Culture Lee Min Sup to Franco Mancuso, written on August 31st, 1994.
©Mancuso e Serena Architetti Associati. Courtesy of ARKO Arts Archive, Arts Council Korea.

At the opening ceremony of the Korean Pavilion.
©Mancuso e Serena Architetti Associati. Courtesy of ARKO Arts Archive, Arts Council Korea.

NAM JUNE PAIK'S BRAVE NEW WORLD: Considering that modern fax technology was developed by Xerox in the late 1960s and only became widely popular worldwide after the late 1970s, the emergence of the pavilion as an organic collaborative design between two countries was made possible by the arrival of this new communication system. From late 1993 until the pavilion's completion, Kim Seok Chul and Franco Mancuso exchanged hundreds of faxes, refining the design and strategizing its realization. Amid this flood of faxes linking Seoul and Venice, one cannot help but recall Nam June Paik's optimistic vision of a new world, where the rise of the information society would blur national borders.

A fax letter sent by Kim Seok Chul to Franco Mancuso on December 10th, 1993.
©Mancuso e Serena Architetti Associati. Courtesy of ARKO Arts Archive, Arts Council Korea.

A fax letter sent by Kim Seok Chul to Franco Mancuso on October 29th, 1994.
©Mancuso e Serena Architetti Associati. Courtesy of ARKO Arts Archive, Arts Council Korea.

⋯ Piazza San Marco is, in the truest sense, an empty space. ⋯ The vast open water where the canal ends flows seamlessly into Piazza San Marco, forming a shared space that belongs to everyone. Above this emptiness, the architecture of a thousand years maintains its own trajectory, creating a grand historical space. ⋯ Piazza San Marco holds nature, history, and, above all, the life of today.

— Kim Seok Chul, *World Architecture Journey* (1997)

INTER-NATIONAL-NESS?: As a product of active joint design, the Korean Pavilion aimed to reflect both Korean and Venetian regional and historical identities. This intention is explicitly confirmed throughout the correspondence between the two architects. Kim Seok Chul, in particular, expressed on multiple occasions his hope that Franco Mancuso, a professor at IUAV — and someone with a deep understanding of Venice — would go beyond the role of a local coordinator or site supervisor and instead actively incorporate Italy's and Venice's historical and regional characteristics into the design.

COMUNE DI VENEZIA
IL SINDACO

     Ribadisco ancora che l'iniziativa è di rilevante interesse pubblico, trattandosi del primo intervento significativo in Italia da parte della Repubblica Sud Coreana, uno dei Paesi dalle maggiori potenzialità industriali, economiche, ma anche culturali, del mondo.

     Cordiali saluti.

Massimo Cacciari

I would also like to emphasize that the initiative is of significant public interest, as it is the first significant intervention in Italy by the South Korean Republic, one of the countries with the greatest industrial, economic, and also cultural potential in the world.

A letter from the Mayor of Venice, Massimo Cacciari, to the president of the Venice Safeguarding Commission, on July 15th, 1994.
©Mancuso e Serena Architetti Associati. Courtesy of ARKO Arts Archive, Arts Council Korea.

DIPLOMATIC JOINT VENTURE: The Korean Pavilion was more than just a design joint venture — it was also a diplomatic joint venture. For the Italian and Venetian authorities, the Korean Pavilion was seen as Korea's first significant intervention in Italy, marking the presence of a rising nation from a new century with notable "potential." At the time, Venice's mayor Massimo Cacciari emphasized that the Korean Pavilion represented the first substantial collaboration between the two nations and therefore carried "public interest." He used this argument to persuade the Venice Safeguarding Commission to approve the project. Cacciari's remarks at the opening ceremony of the Korean Pavilion's introductory exhibition at Spazio Olivetti were even more direct and specific. He stated that the Korean Pavilion would mark Korea's first engagement in Europe as a rising economic power. More strikingly, he argued that it was not Venice choosing Korea, but rather Korea choosing Venice. He even suggested that the pavilion could become a starting point for future Korean business involvement in the redevelopment of Marghera region. Just as the Korean government viewed the pavilion as a device of cultural diplomacy, so too did Cacciari, recognizing its broader geopolitical significance.

Arte, cultura, investimenti a Marghera
Padiglione Corea alla Biennale

Il plastico del padiglione coreano ai Giardini della Biennale

CULTURA e affari all'ombra del Padiglione. E' stato presentato ieri in Comune — alla presenza del ministro coreano della Cultura e dello Sport Lin Min Sup, del sindaco Massimo Cacciari, del segretario generale della Biennale Raffaello Martelli e di una quantità incredibile di giornalisti e operatori televisivi sudcoreani — il nuovo padiglione che la Corea Sud costruirà ai Giardini di Castello, in vista del Centenario di Ca' Giustinian. I lavori dell'opera — che sarà smontabile (visto il permesso teoricamente provvisorio) e costruita su palafitte, senza pregiudicare l'area verde in cui verrà inserita — sono già iniziati ieri e i progetti sono in mostra allo Spazio Olivetti.

Essa sarà realizzata dall'architetto Seok Chul Kim e dal suo collega di architettura Franco Mancuso nello spazio ove oggi sorgono i servizi igienici di tipo ottocentesco, tra il padiglione tedesco e quello giapponese. La spesa prevista per la costruzione è di un miliardo e 400 milioni.

«Vogliamo utilizzare il padiglione», ha detto il ministro coreano, «anche per manifestazioni che favoriscano una reale integrazione tra il nostro Paese e la Corea del Nord, e l'arte e la cultura rappresentano un'ottima base di partenza».

«Per la Corea del Sud, paese in grande espansione economica, questa è la prima uscita europea», ha spiegato Cacciari, «e ha scelto Venezia. Logico che poi pensi anche a investire qui, non solo attraverso il turismo e il commercio. Imprese coreane potrebbero intervenire anche nella riqualificazione di Marghera».

Per questo il Comune ha detto sì alla Corea — che garantisce un'attività continuativa e chiedeva il padiglione dall'86 — scontentando molti altri pretendenti, come il Portogallo, l'Argentina, il Cile. Questi — come altri paesi — dovranno aspettare l'ampliamento dei Giardini verso est, negli spazi ora occupati dai cantieri dell'Actv, perché nell'area attuale ormai non c'è più posto.

"Arte, cultura, investimenti a Marghera - Padiglione Corea alla Biennale"
(Art, culture, investments in Marghera - Korean Pavilion at Biennale), *La Nuova Venezia*, November 4th, 1994.
©Mancuso e Serena Architetti Associati. Courtesy of ARKO Arts Archive, Arts Council Korea

"We want to use the pavilion," said the Korean minister, "also for events that foster real integration between our country and North Korea. Art and culture represent an excellent starting point."

"For South Korea, a country experiencing significant economic expansion, this is its first European appearance," explained Cacciari. "It chose Venice. It is logical that it would then consider investing here, not only through tourism and trade. Korean companies could also intervene in the redevelopment of Marghera."

For this reason, the City has said yes to Korea — which guarantees continuous activity and has requested a pavilion since 1986 — beating out many other contenders such as Portugal, Argentina, and Chile. These countries — like others — will have to wait for the expansion of the Giardini towards the east, in spaces now occupied by ACTV shipyards, because there is no longer room in the current area.

Massimo Cacciari giving a speech at the opening ceremony of the Korean Pavilion of 1995 International Art Exhibition. Photo Giorgio Zucchiatti
Courtesy © Archivio Storico della Biennale di Venezia, ASAC

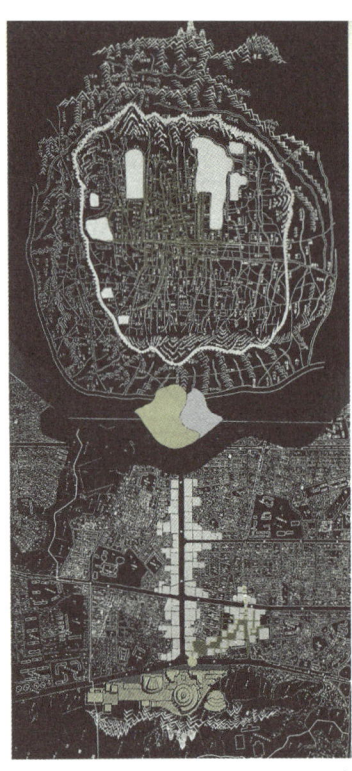

The catalog of Kim Seok Chul's Venice exhibition, *SEOUL, ARCHITETTURA & CITTÀ* (1993).
ⒸKim Seok Chul. Courtesy of ARKO Arts Archive, Arts Council Korea (Contributor: Kim Seok Woo)

THE NEXT KOREAN-NESS?: The connection and friendship between Seoul's Kim Seok Chul and Venice's Franco Mancuso began in 1991. Kim Seok Chul was introduced to Mancuso through Kim Kyung Soo, who was then a visiting professor at IUAV. With Mancuso's support, Kim Seok Chul was invited by IUAV to hold a solo exhibition, SEOUL, ARCHITETTURA & CITTÀ, at Ca' Tron in Venice in 1993. Although the exhibition showcased Kim's architectural and urban design work, its title made it clear that the focus was not on his own architectural style or that of Korea, but rather on the city itself as a theme. For Kim Seok Chul, the expression of Korean-ness was a subject he could not entirely set aside. However, born in 1943, he showed a tendency different from his predecessors. Rather than directly referencing traditional forms, he sought an approach based on more intangible principles, something more fundamental. His interest lay in finding a way to express national identity while integrating it into the universal language of modern architecture.

> The transfer of civilizations began only after the advent of recorded history. That is why I sought to find what is uniquely ours within the relics of the Bronze Age and early Iron Age, eras predating the written word. It was for this reason that, when designing the Korean Pavilion at the Venice Biennale, I presented 'archaeological futurism' as my architectural philosophy.
>
> — Kim Seok Chul, *World Architecture Journey* (1997)

The Korea Times, Jaunary 28th, 1995.
ⒸMancuso e Serena Architetti Associati. Courtesy of ARKO Arts Archive, Arts Council Korea.

"The Korean Pavilion established in the Venice Biennale," *The Daily Sports*, August 25th, 1994.

With the centennial of the Venice Biennale, which has led the world of contemporary art, Korea has become the 25th country to establish a national pavilion in the Giardini park on the outskirts of Venice. This signifies that Korean contemporary art now has a stage where it can stand shoulder to shoulder with the world's leading art nations. Furthermore, the fact that, among 23 countries—including China and Argentina—that sought to build independent pavilions, only Korea was granted approval is a point of pride for our culture.

— "Opening of the Korean Pavilion at the Venice Biennale," *Monthly Culture & Arts*, June 1995.

However, the establishment of the Korean Pavilion was still
a state-led project aimed at enhancing national prestige. By
the mid-1990s, Korea sought to assert its identity on the
international stage, leveraging the rapid economic growth it
had accumulated in the latter half of the 20th century.

```
09-March-94  FRIDAY   16:30    ARCHITECT SEOK CHUL KIM      PAGE 01
ARCHIBAN,  1-132 Dong Soong-Dong Jong Ro-Gu Seoul, Korea
Phone 027643072      Fax 027654651
```

```
Seoul, 09.03.1994
Istituto Universitario Di
Architecttura Di Venezia
To : Prof. Franco Mancuso
Fax : (041) 523-8121
```

Dear Franco,

Thank you for your fax from March 08.
I appreciate your arrangements to meet the mayor of Venice,
the contractor, the architect Feletti and the engineer Ruggero.
I can see our work is progressing.
I met with the Minister of Culture, yesterday.
There are some slight changes in the details for the pavilion.
He would like the pavilion to have a more korean character.
I will bring the new model and the latest AUTOCAD drawings with
me to Venice.
I also hope that we will have enough time for our common work.

Best Ragards and Wishes
Sincerely,

Seok Chul Kim

A fax letter sent by Kim Seok Chul to Franco Mancuso on March 9th, 1994.
©Mancuso e Serena Architetti Associati. Courtesy of ARKO Arts Archive, Arts Council Korea.

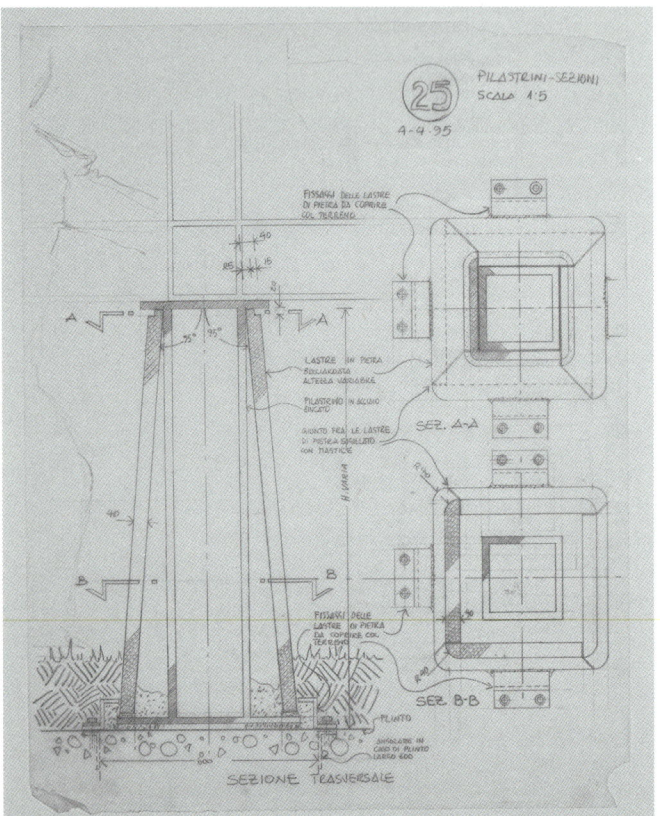

©Mancuso e Serena Architetti Associati. Courtesy of ARKO Arts Archive, Arts Council Korea.

VESTIGIAL ORGANS: As described by contemporary Korean media at the time of its completion, the Korean Pavilion appeared futuristic, resembling a "spaceship." However, much like vestigial organs that persist from the past, certain physical elements derived from traditional Korean architecture remained embedded in its design. For instance, while the pavilion incorporated what was then the most advanced kinetic security window system among the Giardini pavilions, its grilles directly referenced traditional wooden lattice patterns and was installed in a manner reminiscent of Korean traditional "deureo-yeolgae-mun" (hinged folding doors). The elevated structural base, designed to preserve the natural terrain, also consciously reflected traditional forms. This was particularly evident in the choice to clad the steel piles with bush-hammered stone slabs, evoking the use of granite in traditional Korean architecture. Later, the wooden lattice were replaced with steel, gradually causing the traces of tradition to fade out.

The opening ceremony of the Korean Pavilion (1995)
©Mancuso e Serena Architetti Associati. Courtesy of ARKO Arts Archive, Arts Council Korea.

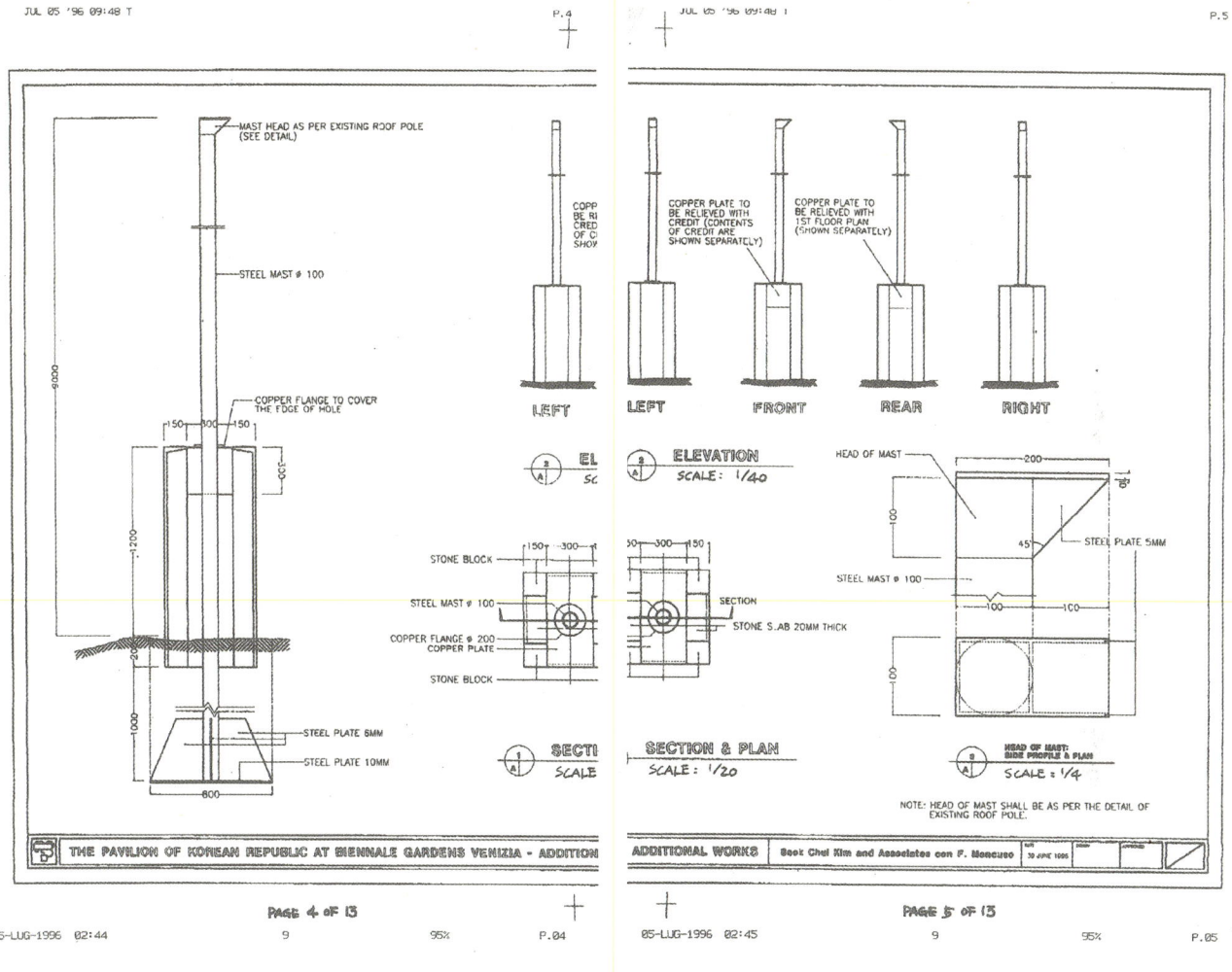

Unrealized 9 meter-height flagpole, planned to be installed in front of the pavilion, following the request of the Korean Culture and Art Foundation (the former entity of Arts Council Korea).
©Mancuso e Serena Architetti Associati. Courtesy of ARKO Arts Archive, Arts Council Korea.

Proposal for the Korean Pavilion

Interpreting the Land

The site is located inside the Giardini della Biennale, behind the Russian, German and Japanese pavilions. It is on a hillside between the existing pavilions and the lagoon in the north-west. There are trees and an existing restroom on the site.

The approach happens from inbetween the Japanese and the German pavilions, continuing to flow from the existing pavilions to the site.

Transparency between the elements of the structure is suggested for the site.

Understanding the Theme

The pavilions stay dormant for most of the time, when there is no Biennale or other festival.

The Korean Pavilion is added among the existing pavilions in the Giardini della Biennale, which have been standing there for a long time, belonging to the 100 year history of the place.

The existing restroom on the site is one of the historical buildings in the Giardini della Biennale. Integrating the existing restroom to the new pavilion, by enlarging the restroom and creating a new public space, would include the new pavilion into the flow of historical memories. However, the restroom still stays as an individual building in the Giardini Pubblici.

The new pavilion is transparent, so the existing pavilions can be seen from the lagoon side. It is integrated into the natural environment, maintaining the existing conditions of the site.

The new pavilion reflects an archeological image when there is no festival. I hope that your contribution will help our project to become more sympathetic to the site and the culture of Venice.

At the time of a Biennale this dormant pavilion erupts as an instant exhibitional hall, for modern art and expression to be visually seen inside and outside of the pavilion. It shows the sudden change of expressiveness giving an abstract awareness.

The new pavilion will then be a temporary exhibition space that is highly alive. However, once the festival is over, it will return to its original state of dormancy.

Proposal   PAGE 03

The structure of the new pavilion consists of 12 columns, integrated into the existing natural environment of trees and the existing restroom building on the site.

Each column is composed of 5 elements of different materials. The pile is made of steel, the pile cap, wrought iron, the shaft out of wood and iron, the capital out of Bronze and crowning the column would be a lantern out of Murano Glass.

The pile caps and capitals are constructed such that they can be connected with separate parts, creating a floor, a roof, ceiling and walls for an "instant exhibition hall", when there is a Biennale. When there is no festival, it would be a skeletal or transparent structure, preserving the natural atmosphere and historical beauty of place.

The walls of the instant exhibition hall are used from both sides, holding exhibitions inside and outside. There would be no specific distinction between the indoor and outdoor space. Canopies on the outside, would give cover from rain and would be used for temporary lighting installments.

The existing restroom is integrated into the 12 columns. A new area would be created, that becomes the rest area with the existing restroom building.

The informal structure of the building does not block the view from the lagoon towards the existing pavilions. The new pavilion would preserve the beauty of the natural environment of trees. " It would be a place in itself."

These 12 columns provide an abstract structure. These 12 columns have an archeological image.

The initial proposal for the Korean Pavilion by Kim Seok Chul, sent to Franco Mancuso on October 18th, 1993.
©Mancuso e Serena Architetti Associati. Courtesy of ARKO Arts Archive, Arts Council Korea.

Project Description

The project begins from not destroying the existing site of the Giardini de la Biennale.

The site is located in the backyard of the main street. It looks like a deserted area. However, we find great possibilities for it to become a new attractive scenery. The quietness of the site can be converted into the merit that korean traditional architecture has in common. The ultimate goal which we try to establish in this project is, to represent the spirit which korean traditional architecture has preserved.

The basic concept of the structure is transparency. This transparency supports the frequent communication between the outside and the inside. The basic idea about how the pavilion is used, comes from how the korean traditional festival has been held. Korean people gathered every five or six days in which the gathering became a natural market. The pavilion is located on the street so that in common days it is used as a secret and quiet memorial-like building whereas in special days it will be the strong supporter for maximizing the happiness and surprise.

Visitors approach the structure from the level between the japanese and the german pavilions. Once inside the pavilion, you are under the level of the hilltop. The structure is lowered 2.5 meters into the hill, with respects to the sea level. In the inside are 4 columns. A sliding canvas roof in the form of an arch is installed when there is a festival, to establish a temporary exhibition hall. When there is no festival, the roof is taken down and it is a garden structure.

The walls are permanent structures, but not visible from the level of the hill due to being partly underground and transparent glass structures with an overhang above the ground. When there is no festival, the pavilion is a back garden for people to enjoy, walk through and rest. A glass window with a view towards the lagoon beautifies the garden structure. This pavilion is a circulating garden in organic form that is transparent from the ground level of the hill.

The existing restroom on the site is restored for comfortable use and one part of the new pavilion is a new additional restroom space for showers.

Finally, we expect this project will be a contribution to a co-understanding between the italian and korean culture, especially in architecture.

The proposal for the Korean Pavilion by Kim Seok Chul, sent to Franco Mancuso on November 9th, 1993.
©Mancuso e Serena Architetti Associati. Courtesy of ARKO Arts Archive, Arts Council Korea.

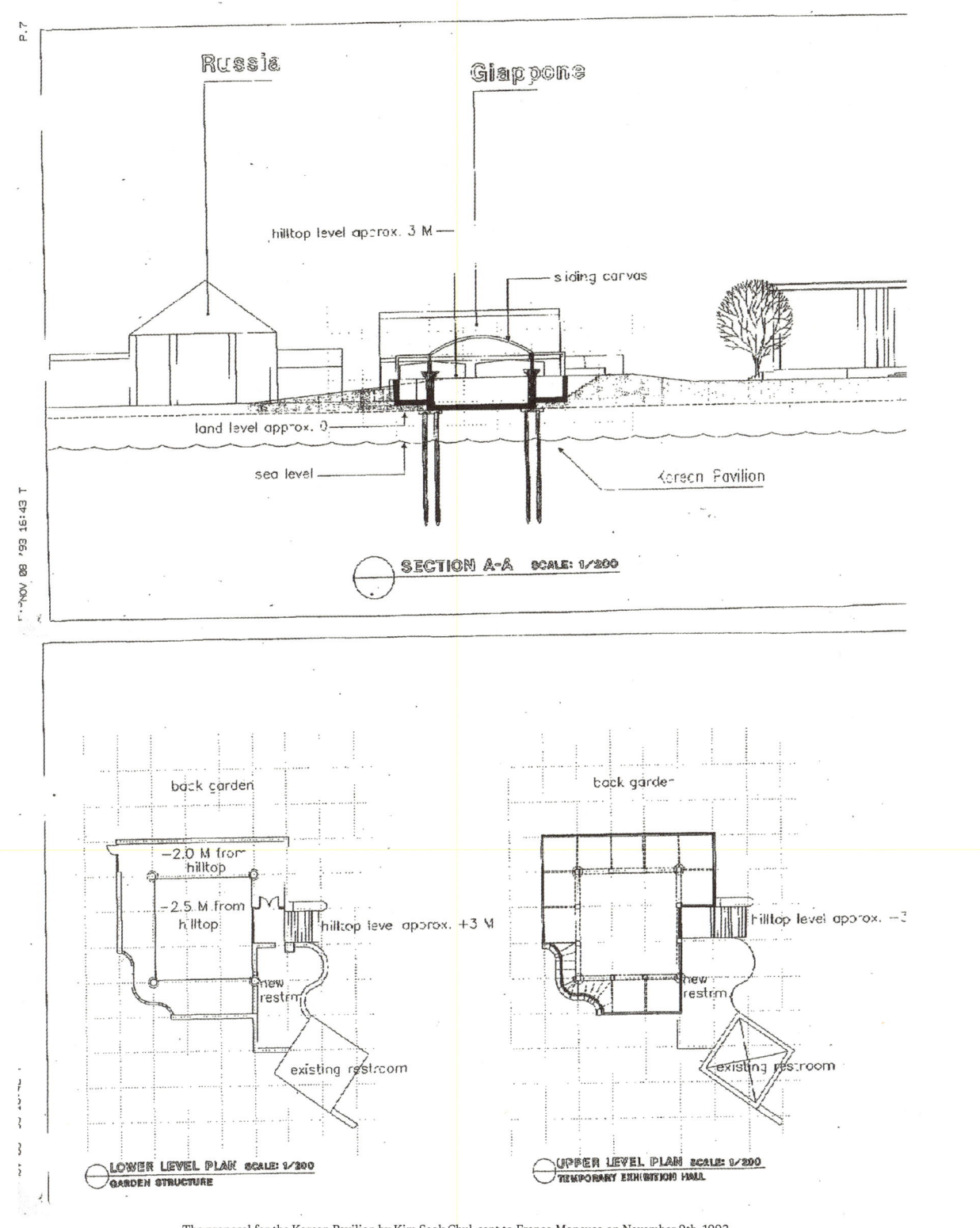

The proposal for the Korean Pavilion by Kim Seok Chul, sent to Franco Mancuso on November 9th, 1993.
In the early design stages, the concept of visual transparency was so crucial that the two architects even discussed an underground design,
where the entire structure would be buried below ground, with a temporary sliding canvas roof installed only during the biennale.
©Mancuso e Serena Architetti Associati. Courtesy of ARKO Arts Archive, Arts Council Korea.

TRANSPARENCY AS A DESTINED TACTIC: The core architectural identity of the Korean Pavilion lay elsewhere. Amid the Biennale and Giardini's longstanding closedness and the shifting atmosphere of openness marking the centennial transformation, the pavilion — an international joint venture — adopted a timely tactic, navigating a delicate balancing act: The proposal of a new type of pavilion that functioned both as a national representation and an open public space. The architectural response to this paradox was "transparency."

Transparency first fulfilled the expression of Korean-ness in terms of harmony with nature, without resorting to direct formal citations of traditional architecture. More critically, however, it provided the necessary justification for adding a new pavilion to the already packed Giardini at a moment when the Biennale and Giardini were opening their doors to a new century. Through "disappearing transparency," the Korean Pavilion was designed to ensure a physical premise that would not compromise the Giardini's publicness even during the periods when the Biennale was not in session. This approach envisioned the pavilion as a non-fixed, open space that would seamlessly revert to being part of the park once the festival ended. This concept aligned with the temporariness of Korea's traditional markets, further reinforcing Korean-ness. The pavilion was designed less for its function as an exhibition hall and more for its ability to dissolve into the park as a public space.

Within the broader transformation of both the Biennale and the Giardini, this transparency was an architectural inevitability — the only viable form the Korean Pavilion could take in order to be realized. In other words, while the post-Cold War geopolitical landscape and the vision of a joint North and South Korean pavilion made the pavilion's establishment possible, it was the conjunction of emerging discourse on publicness and the desire of a non-physical national expression that determined the shape it was built.

TRANSPARENCY BORN FROM EXTREMES: From its very first design ideas, the Korean Pavilion's intention to both represent a national identity and disappear as part of the park is clearly evident in the surviving archival materials. Consider the skeletal structure of the initial design concept: a stripped-down form reduced to nothing but a colonnade of twelve pillars. Named the "Temporary Exhibition Hall," this structure presents the most fundamental and extreme prototype of the transparency and lightness that define the Korean Pavilion today.

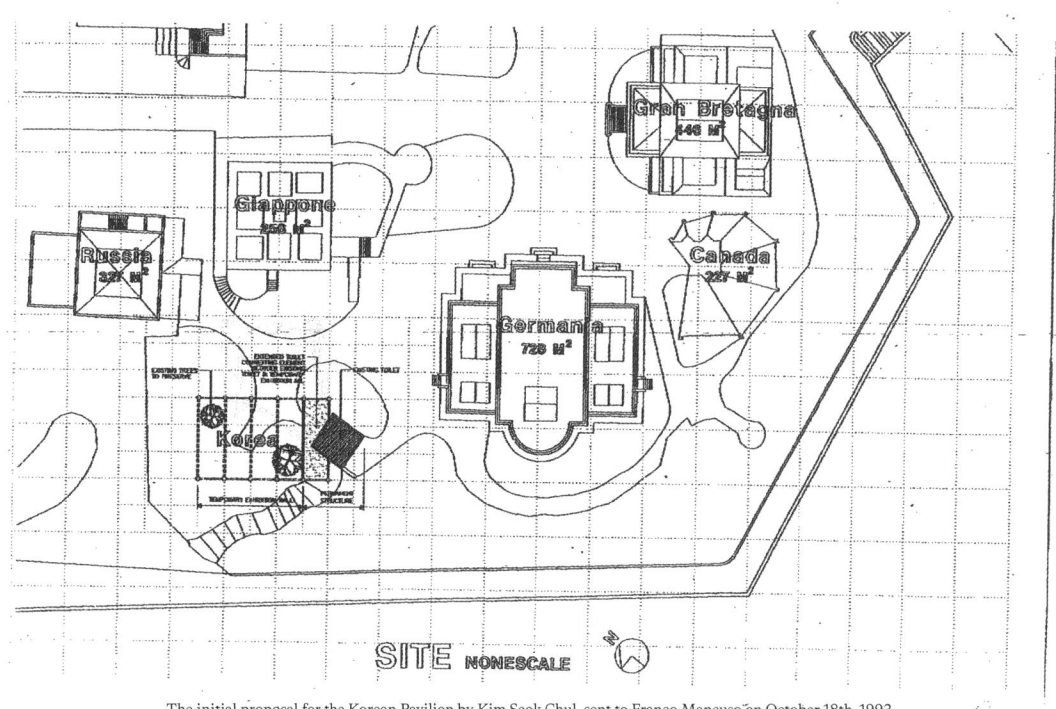

The initial proposal for the Korean Pavilion by Kim Seok Chul, sent to Franco Mancuso on October 18th, 1993.
©Mancuso e Serena Architetti Associati. Courtesy of ARKO Arts Archive, Arts Council Korea.

TRANSPARENT GRID: In the early design phase, the language of a transparent grid was emerged by embedding the most archaeological form of a colonnade into the ground to seamlessly blend into the environment. This planar principle remained a fundamental axiom throughout the design process and was ultimately materialized as a lightweight and dismantlable steel structure.

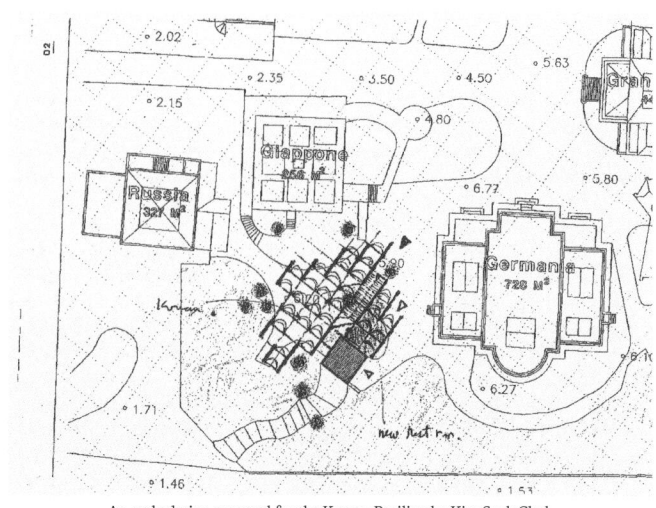

An early design proposal for the Korean Pavilion by Kim Seok Chul,
sent to Franco Mancuso on November 11th, 1993.
©Mancuso e Serena Architetti Associati. Courtesy of ARKO Arts Archive, Arts Council Korea.

TO: 김석철 소장님
FROM: 박창근, VENICE, 12:30(PM : 현지시간), 5/9

- 오전에 모두 모여서 (만쿠조 사무실) 최종 계약서에 합의 했습니다. 내역서 확인하고 계약서 문구 수정하고 어버린 서류를 청구하느라고 시간이 좀 걸렸습니다. 내일 (5/10) 최종계약서 (ICCEN사의 sign이 첨부된)를 받아서 서울로 가져가겠습니다.

- 어제 오후에 ICRA에 들렀습니다. 모든 물량이 거의 완성 되어서 현장에서의 요청을 기다리는 상태이고 (15日경이면 모든 작업이 마무리 된다고 합니다) 제작상태는 만족스럽습니다. 유리는 모두 완료되어 현장에 설치되었고 창광에서는 Cylinder의 음속파넬과 천정, 바닥 (화강석)의 음속판을 보았습니다. 어제는 시간상 ICRAS만을 보았고, 오늘 오후 NAUTICA에 들러서 WOODEN WALL을 보려고 합니다.

- 건물에 'COREA' 글자를 부착하는 문제와 국기게양대를 마련하는 문제: 글자는 일부 BEAM에 매다는 형식으로 하고 국기게양대는 현재 두께의 POST를 이용하는 방법으로 MANCUSO와 합의하였습니다. 자세한 내용은 서울에 돌아가서 말씀드리겠습니다. 현관 CANOPY는 MANCUSO와 협의했는데 각자 더 생각해보고 내일 의논하기로 했습니다. CANOPY를 제작하는 시간상의 문제도 있긴 합니다. 전수천씨 작품은 만쿠조교수가 Brutomeso이와 함께 현지를 한장 써주기로 하였습니다.

- 현장사진이 도착했는지 궁금합니다. 한국관은 TRANSPARENT 하면서도 Object로서의 성격이 주위의 다른판에 비해 매우 강합니다. 쟈르디니에서 가장 독특한 건물이 될 것 같습니다. 건물의 페인트색(내부및 외부)은 잘 선정되었다고 생각합니다. 지붕위의 POST와 난간 디테일로 약간의 아쉬움이 남습니다 (처음의 안이 조금더 보존됐으면…)

다시 연락드리겠습니다.

A fax letter from Park Chang Geun, working at the Korean Pavilion construction site in Venice, to Kim Seok Chul, May 9th, 1995.
©Mancuso e Serena Architetti Associati. Courtesy of ARKO Arts Archive, Arts Council Korea.

> I am wondering if the site photos have arrived. The Korean Pavilion is transparent, and a character as an object is much stronger compared to the surrounding pavilions. It will be the most distinctive building in Giardini. … The post and railing details on the roof leave some regrets. (I wish the initial design had been preserved a little more…)

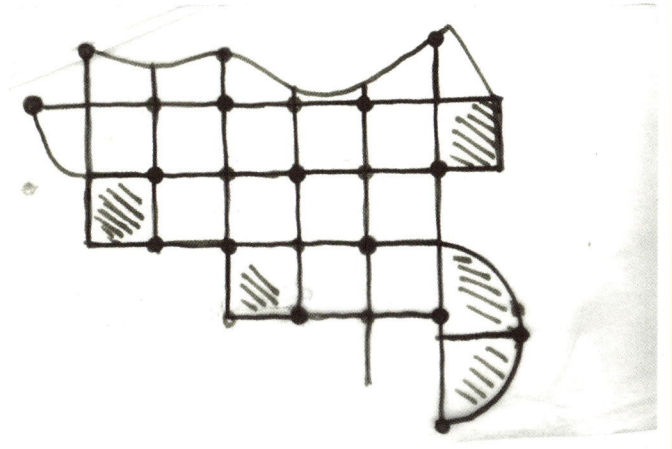

Franco Mancuso's conceptual floor plan sketch of the Korean Pavilion. The composition of the grid is distinctly visible.
©Mancuso e Serena Architetti Associati. Courtesy of ARKO Arts Archive, Arts Council Korea.

The structural floor plan included in the set of drawings created by Franco Mancuso in April 1994. In a form close to the final design, the grid, initially conceived as a design concept, was materialized into a steel frame structure.
©Mancuso e Serena Architetti Associati. Courtesy of ARKO Arts Archive, Arts Council Korea.

Gill, potresti tradurre al più presto (oggi) queste lettere? 5240295
grave molto

Venezia, 10.05.1995

To: Arch. Seok Chul Kim
ARCHIBAN
Seoul (Korea)

fax 0082.2.765.4651

Caro Seok Chul,

sabato 6 Maggio, durante il sopralluogo al cantiere (c'erano anche Mr. Park e i funzionari della Korean Foundation), il Padiglione Koreano ha ospitato la visita di una delegazione ufficiale della Biennale, guidata dall'avvocato Martelli (segretario Generale) e dal prof. Romanelli (Direttore dei Musei Civici).

Tutti i membri della Delegazione hanno espresso le loro congratulazioni alla Korea, per il nuovo padiglione, apprezzandone i caratteri fondamentali, e soprattutto la trasparenza e l'affaccio sul Bacino S. Marco. Tutti si aspettano di vederlo finito, e di poter finalmente godere di queste indiscutibili qualità.

Di fronte a questa aspettativa, che è anche quella del Sindaco e di tutti coloro che hanno contribuito al successo dell'iniziativa, non possiamo non comunicarti il nostro sincero invito ad operare affinchè la sistemazione delle opere d'arte non comprometta le doti essenziali del Padiglione -trasparenza e affaccio sul Bacino- che sono quelle stesse per le quali il progetto era stato a suo tempo approvato dalla Commissione di Salvaguardia e dal Comune.

A tutti noi sta a cuore, come a voi, che lo sforzo della Korea abbia il massimo del successo: ma abbiamo la netta sensazione, dopo la visita della Delegazione della Biennale, che l'allestimento corra il rischio di essere fortemente criticato, se non valorizza le doti intrinseche del Padiglione, che tutti si aspettano di vedere esaltate durante l'esposizione. E a noi sembra che l'installazione prevista nella sala principale -con l'allestimento di un ambiente cieco che chiude tutte le pareti vetrate, e impedisce l'accesso alle balconate- sia estremamente controproducente.

C'è poi un altro problema, che desideriamo segnalarti. Abbiamo la netta sensazione che l'allestimento proposto per la sala principale interferisca pesantemente con gli impianti tecnici: con il sistema di illuminazione, e in particolare con l'impianto di condizionamento -essenziale durante il periodo estivo- che verebbe intercettato dalle pareti cieche previste dall'allestimento.

Ti preghiamo di considerare con attenzione le nostre preoccupazioni, che sono dettate unicamente dal desiderio che la presenza della Korea alla Biennale di Venezia abbia il massimo del successo.

Cari saluti.

Franco Mancuso
Rinio Bruttomesso

A fax letter sent by Franco Mancuso and Rinio Bruttomesso to Kim Seok Chul on May 10th, 1995.
©Mancuso e Serena Architetti Associati. Courtesy of ARKO Arts Archive, Arts Council Korea.

On Saturday, May 6, during an inspection of the construction site (where Mr. Park and representatives of the Korean Foundation were also present), the Korean Pavilion received a visit from the official Biennale delegation. This delegation was led by Lawyer Martelli (Secretary General) and Professor Romanelli (Director of Civic Museums).

All members of the delegation expressed their congratulations on Korea's new pavilion, appreciating its basic characteristics and especially its transparency and view of the San Marco Basin. Everyone is looking forward to seeing the pavilion completed and enjoying these undoubted qualities.

In response to these expectations, like the Mayor and all those who have contributed to the success of this project, we ask that you ensure that the placement of artworks does not compromise the essential elements of the pavilion's transparency and view of the San Marco Basin. These are the same elements for which the project has already been approved by the Protection Commission and the City Hall.

Like you, we sincerely hope that Korea's efforts will be as successful as possible. However, following the delegation's visit, we felt there was a risk that the installation might be criticized at this Biennale or might not sufficiently enhance the pavilion's inherent characteristics. We are concerned that it might not emphasize the pavilion's intrinsic qualities that everyone expects to see during the exhibition. In particular, we are very worried about the closed structure installed on the wall towards the main balcony, which isolates it from the surroundings, obstructs the view, and makes access to the balcony difficult.

FORESEEN COLLISION: Indeed, the Biennale authorities also recognized transparency and its harmonious integration with the surrounding nature as the defining value of the Korean Pavilion, given its fundamentally different approach from the existing national pavilions in the Giardini. Rinio Bruttomesso, an IUAV professor and the local coordinator who welcomed the Biennale delegation on the construction site, sensed that transparency would ultimately be regarded as the pavilion's most significant characteristic. Anticipating this, he expressed early concerns that the installation for the 1995 International Art Exhibition, which would revert the pavilion into a white or black cube through various temporary partitions, might face criticism.

©Mancuso e Serena Architetti Associati. Courtesy of ARKO Arts Archive, Arts Council Korea.

```
MAY 12 '94 18:17 T                                                    P.2

        Proposal for the Program at the Korean Pavilion, Venice Biennale


     - The Korean Pavilion will open every two years during the Biennale
       Exhibition.  There will be a 1 month preparation time before the
       3 month Biennale Exhibition, alltogether 4 months.  The pavilion will
       exhibit korean artifacts during these four months at the Biennale.

     - The Korean Pavilion can be used as an entrance or passage during the
       events of the Biennale by the people of the event and the citizens of
       Italy.  The pavilion can be air-conditioned all year around
       for comfortable use.

     - During the time of the Biennale (every 2 years for 4 months), the
       Korean Pavilion is hoped to be a Cultural Space as well as at the
       "Carnevale, Vogalonga, Redentore, Regatta and Salute" for the
       the people of Venice and the citizens of Italy.

     - When the Korean Pavilion is not used for cultural purposes, the
       government of Venice, the people of Venice and the citizens of Italy are
       welcome to use it for different purposes.

     - When the government of Venice does not use the Korean Pavilion,
       it can be used for advertising and awareness purposes of the korean
       culture to the citizens of Italy.
       Pre-contact with the Venice Government or the Venice Biennale
       Commitee should be made.
       The Korean Pavilion represents the opportunities to absorb another
       culture by the people of Venice and the citizens of Italy.
       The following events are possible:

     - Diverse Exhibitional Events:
         * korean design, crafts, fashion, photographs, costumes, arts.
         * korean culture and traditional arts exhibitions.
         * exhibition of works from famous korean artists all over the world.

     - Performance Events;
         * korean music, traditional dance, modern dance, plays.
         * korean movies, periodic video projections in intervals.

     - Korean Lecture of korean history and culture.

     - korean books for exhibits or for library purposes.
```

Proposal for the public program at the Korean Pavilion, written by Ministry of Culture, Korea, translated into English by ARCHIBAN (May 1994)
©Mancuso e Serena Architetti Associati. Courtesy of ARKO Arts Archive, Arts Council Korea.

MINISTER OF CULTURE AND SPORTS
SEOUL, KOREA

Venezia, 05.09.1994

Egr. Prof. Massimo Cacciari
Sindaco del Comune di
VENEZIA

Egregio Signor Sindaco,

desidero anzitutto ringraziarla, a nome del Governo Coreano, per il suo personale interessamento a favore della realizzazione del Padiglione della Repubblica di Corea ai Giardini di Castello.

Il rilascio dell'autorizzazione ci consente di passare ora alla fase più propriamente operativa, che ci ripromettiamo di condurre nel rispetto delle scadenze connesse alla celebrazione del Centenario della Biennale.

Desidero con l'occasione riconfermarle gli impegni che il Governo della Repubblica di Corea intende assumere, considerando la realizzazione del Padiglione come il momento iniziale di un rapporto più ampio con la città di Venezia. In questa prospettiva, accettando senza riserve tutte le condizioni poste nell'autorizzazione stessa, il Governo si impegna in particolare:
- ad avviare concretamente un programma di iniziative culturali connesso ad un uso del Padiglione anche in periodi diversi da quelli interessati dalle manifestazioni della Biennale;
- a tenere aperto il Padiglione durante tali periodi, mettendolo a disposizione della città di Venezia per le attività e le iniziative di carattere culturale legate al programma formulato d'accordo con la Municipalità;
- a realizzare, con gli stessi tempi necessari per la costruzione del Padiglione, le strutture per i servizi igienici pubblici sostitutivi, nelle forme, dimensioni e localizzazioni che gli Uffici competenti del Comune vorranno indicare.

Desidero infine comunicarle che, al di la di questi impegni che formalmente il Governo della Repubblica di Corea ha assunto, il nostro

Paese, nella persona del Ministro della Cultura, è aperto ad ogni altro suggerimento o richiesta che la città di Venezia vorrà avanzare, al fine di sviluppare su un più ampio ventaglio di iniziative i nostri rapporti.

Ringraziandola ancora, le porgo i miei migliori saluti.

Woong Ho, Lee

Lee Woong Ho
(Direttore Generale, Ministero della Cultura)

MINISTER OF CULTURE AND SPORTS
SEOUL, KOREA

Venice, 05.09.1994

To:
Prof. Massimo Cacciari
Mayor of the Municipality of VENICE

Dear Mr. Mayor,
I would like to first thank you, on behalf of the Korean Government, for your personal interest in supporting the realization of the Pavilion of the Republic of Korea at the Giardini di Castello.
The granting of authorization now allows us to move to a more operational phase, which we are committed to carrying out in compliance with the deadlines connected to the celebration of the Centennial of the Biennale.
On this occasion, I would like to reaffirm the commitments that the Government of the Republic of Korea intends to undertake, considering the realization of the Pavilion as the initial moment of a broader relationship with the city of Venice. In this perspective, by unreservedly accepting all conditions set forth in the authorization itself, the Government commits to:

- concretely initiating a program of cultural initiatives connected to the use of the Pavilion even during periods other than those covered by Biennale events;
- keeping the Pavilion open during these periods, making it available to the city of Venice for activities and initiatives of a cultural nature linked to a program formulated in agreement with the Municipality;
- constructing, within the same timeframe necessary for building the Pavilion, replacement structures for public restroom facilities, in forms, dimensions, and locations that will be indicated by competent municipal offices.

Finally, I wish to inform you that, beyond these commitments undertaken formally by the Government of the Republic of Korea, our country — through its Minister of Culture — is open to any other suggestions or requests that the city of Venice wishes to propose in order to develop our relations through a broader range of initiatives.
Once again thanking you, I extend my best regards.

Sincerely,
Lee Woong Ho
(General Director, Ministry of Culture)

FROM PHYSICAL PUBLICNESS TO PROGRAMMATIC PUBLICNESS: Based on the concept of a transparent pavilion that forms an open, non-fixed space, the design process of the Korean Pavilion gradually evolved into a broader brainstorming effort on the diverse public programs that could take place there — particularly during the months when the Biennale was not in session. The final stage proposal submitted to the city of Venice under the name of the Ministry of Culture and Sports of Korea, included programmatic suggestions on how the space could be used outside the four months designated for Biennale preparation and exhibition periods. The proposal suggested that the Korean Pavilion could serve as a venue for local Venetian festivals such as "Carnevale, Vogalonga, Redentore, Regatta, and Salute." Additionally, while the city of Venice and the Italian authorities were encouraged to suggest their own uses, it was also intended to host exhibitions, performances, events, lectures, and book fairs related to Korean culture. In a letter sent to Mayor Massimo Cacciari in September 1994, Minister of Culture General Director Lee Woong Ho reaffirmed that the pavilion had been designed to remain open beyond the Biennale period. He further welcomed the idea of Venice actively proposing additional public programs, reinforcing the pavilion's role as a shared cultural space.

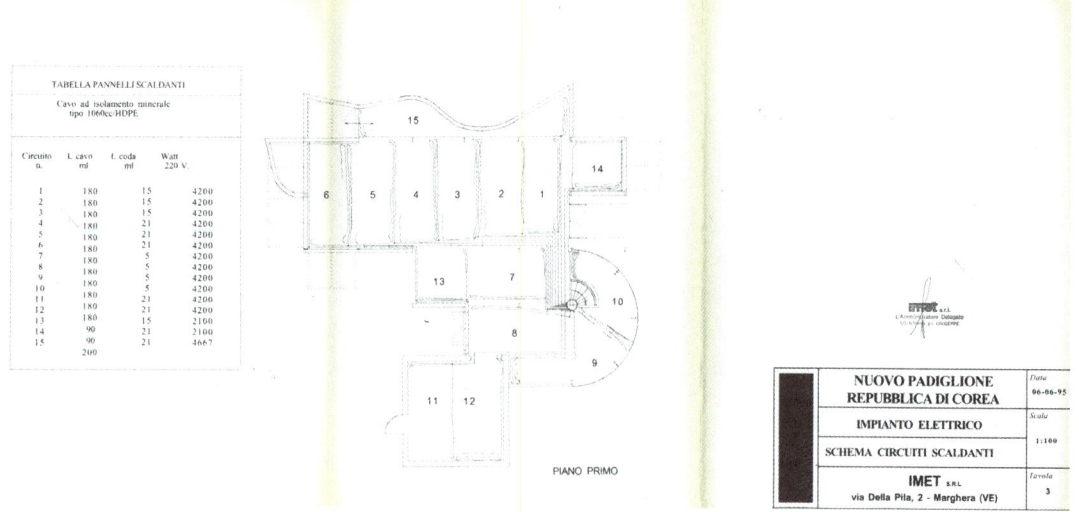

Heating circuit diagram of electrical system for the Korean Pavilion, describing fourteen floor heating panels, drawn by IMET
ASAC, Fondo storico, Progetti speciali b. 5874
Courtesy © Archivio Storico della Biennale di Venezia, ASAC

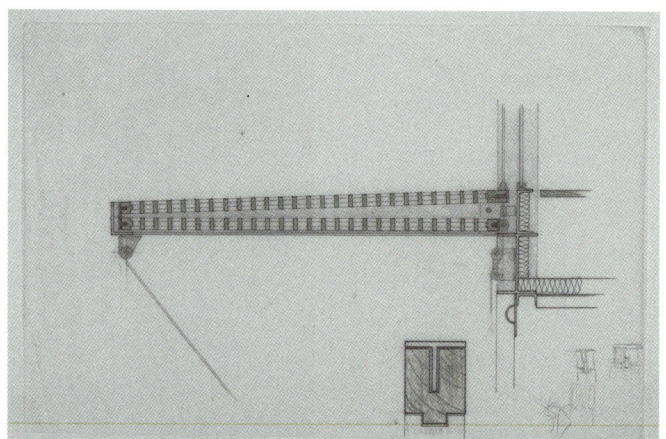

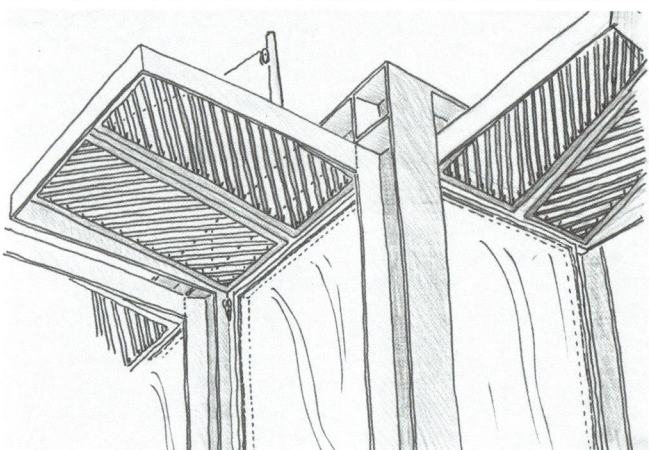

Construction drawings for the kinetic window grilles.
©Mancuso e Serena Architetti Associati. Courtesy of ARKO Arts Archive, Arts Council Korea.

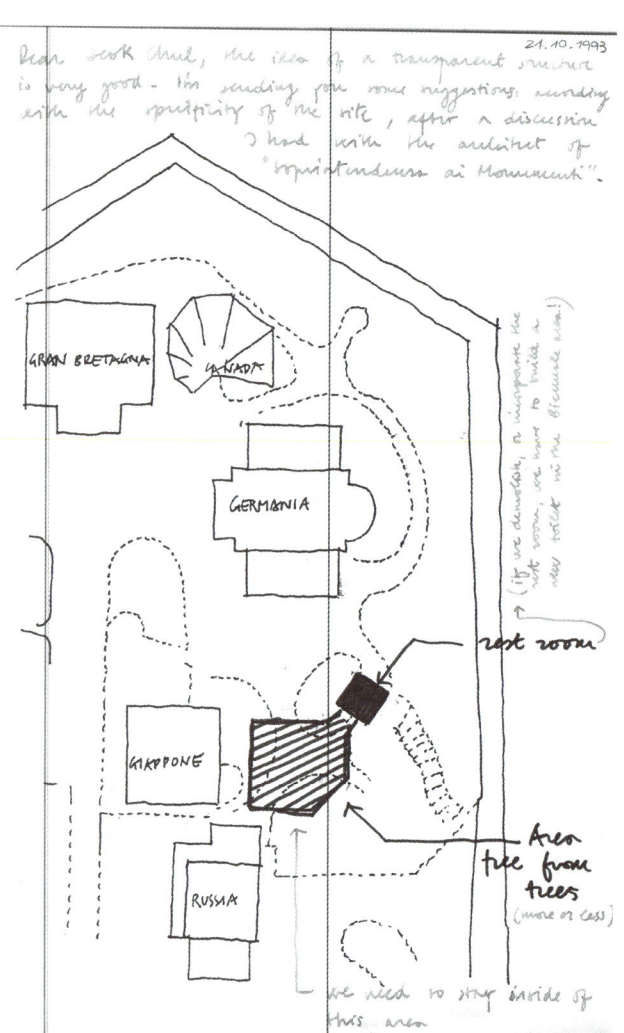

Response of Franco Mancuso on the initial proposal of Kim Seok Chul, drawn on October 21st, 1993.
©Mancuso e Serena Architetti Associati. Courtesy of ARKO Arts Archive, Arts Council Korea.

BECOMING A "HOUSE" FOR ALL SEASONS: For the Korean Pavilion to function as an everyday space where citizens could freely come and go, and where public programs could take place for extended periods beyond the Biennale exhibitions, it needed to ensure a sense of livability. To achieve this, the pavilion was equipped not only with an air conditioning system but also with floor heating, making it the only national pavilion in Giardini to incorporate such a feature. This was a natural architectural choice for Korea, where the tradition of "ondol" (floor radiant heating) remains an integral element of residential spaces today. Additionally, the pavilion's window grilles were installed with an automated kinetic system, allowing users to adjust the amount of natural light as easily as drawing curtains in a home. This feature ensured a balance between constant openness and practical security. Even today, many pavilions in the Giardini install bulky, makeshift protective barriers during the off-season.

Dear Seok Chul, the idea of a transparent structure is very good. I'm sending you some suggestions, according with the specificity of the site, after a discussion I had with the architect of "Soprintendenza ai Monumenti[Supervision of Monuments]."
…
restroom (if we demolish, or incorporate the restroom, we have to build a new toilet in the Biennale area!)

The brick building before the construction of the Korean Pavilion
©Mancuso e Serena Architetti Associati. Courtesy of ARKO Arts Archive, Arts Council Korea.

THE PAST OF THE GIARDINI STANDING ON THE LAND OF THE FUTURE: As seen in the early exchanges of ideas between Kim Seok Chul and Franco Mancuso, the two defining conditions of the Korean Pavilion site were nature and history. The most immediate historical reference they needed to respect was a brick restroom already present on the site.

"The existing restroom on the site is one of the historical buildings in the Giardini della Biennale. ⋯ preserving the natural atmosphere and historical beauty of the place." (Kim Seok Chul, October 18, 1993)

The location of this structure carries an ironic significance. At the 1993 International Art Exhibition, the first post-Cold War Biennale and a prelude to the next century, Nam June Paik scattered futuristically reinterpreted historical figures across this very site. Yet, on that same ground stood one of Giardini's most humble and archaic remnants. The Korean Pavilion project inherited this re-territorialized land, finding itself in a unique position. It was going to be the first national pavilion of a new era, yet at the same time, it had to directly confront the Giardini's historical legacy.

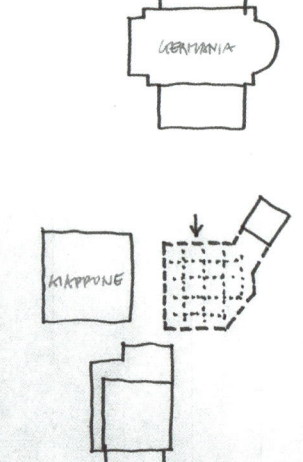

Conceptual sketch of Franco Mancuso as a response to the initial proposal of Kim Seok Chul, sent to Kim on 21, October, 1993.
©Mancuso e Serena Architetti Associati. Courtesy of ARKO Arts Archive, Arts Council Korea.

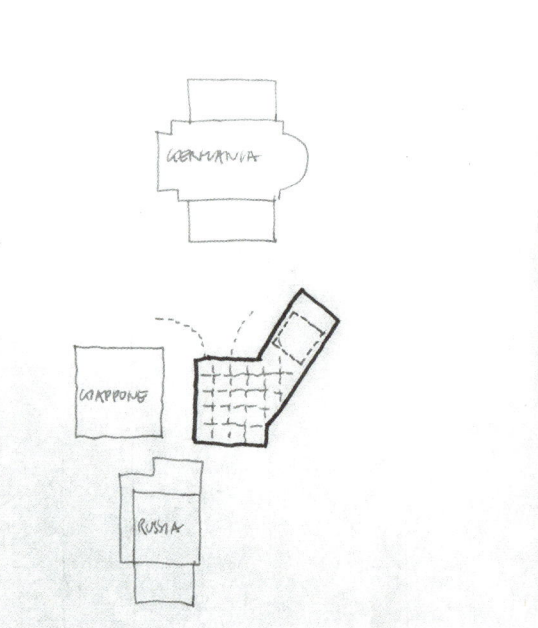

SOLUTION A: to build a transparent new structure, incorporating the existing rest room. The shape of the pavilion is more or less that of the area free from trees - A shape not too geometrical, but organic, according with the gardening of the area.

SOLUTION B: The rest room and a first part of the pavilion (transparent structure) are the entrance, on the top of the hill. Here there is a ramp, and the second part of the pavilion is underground, with top lights. The exit is at the end of the main avenue between Russia and Giappone.

SOLUTION C: We get the permission to demolish the rest room. The new pavilion is on the place of the rest room, with a part on the area free from trees. The shape of the transparent structure follows that of the available area.

In the early stages of design, Kim Seok Chul and Franco Mancuso explored different options based on how they could engage with the brick restroom building and discussed various possibilities.

Map of the Giardini (1919)

AN OLD WITNESS OF THE GIARDINI: The brick building is believed to have been constructed in the 1930s, but it already appears in a smaller form on the 1919 map of Giardini. At that time, only seven nations, aside from Italy, had been invited into the park: the Netherlands, Belgium, Spain, Hungary, Russia, the United Kingdom, and France. Over the decades, this brick structure endured as an office for the Venice Biennale authorities and a public restroom, witnessing the transformation of the Giardini.

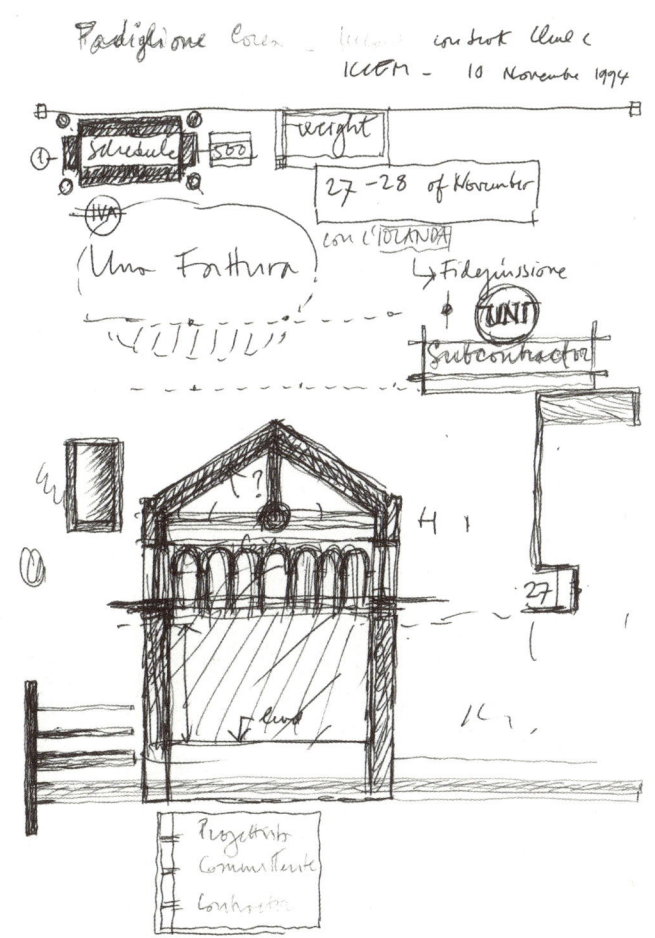

Franco Mancuso's sketch on the brick building. Reads: "Una Forttura (A Fort)"
©Mancuso e Serena Architetti Associati. Courtesy of ARKO Arts Archive, Arts Council Korea.

Construction scene of the Korean Pavilion, preserving and restoring the brick building
©Mancuso e Serena Architetti Associati. Courtesy of ARKO Arts Archive, Arts Council Korea.

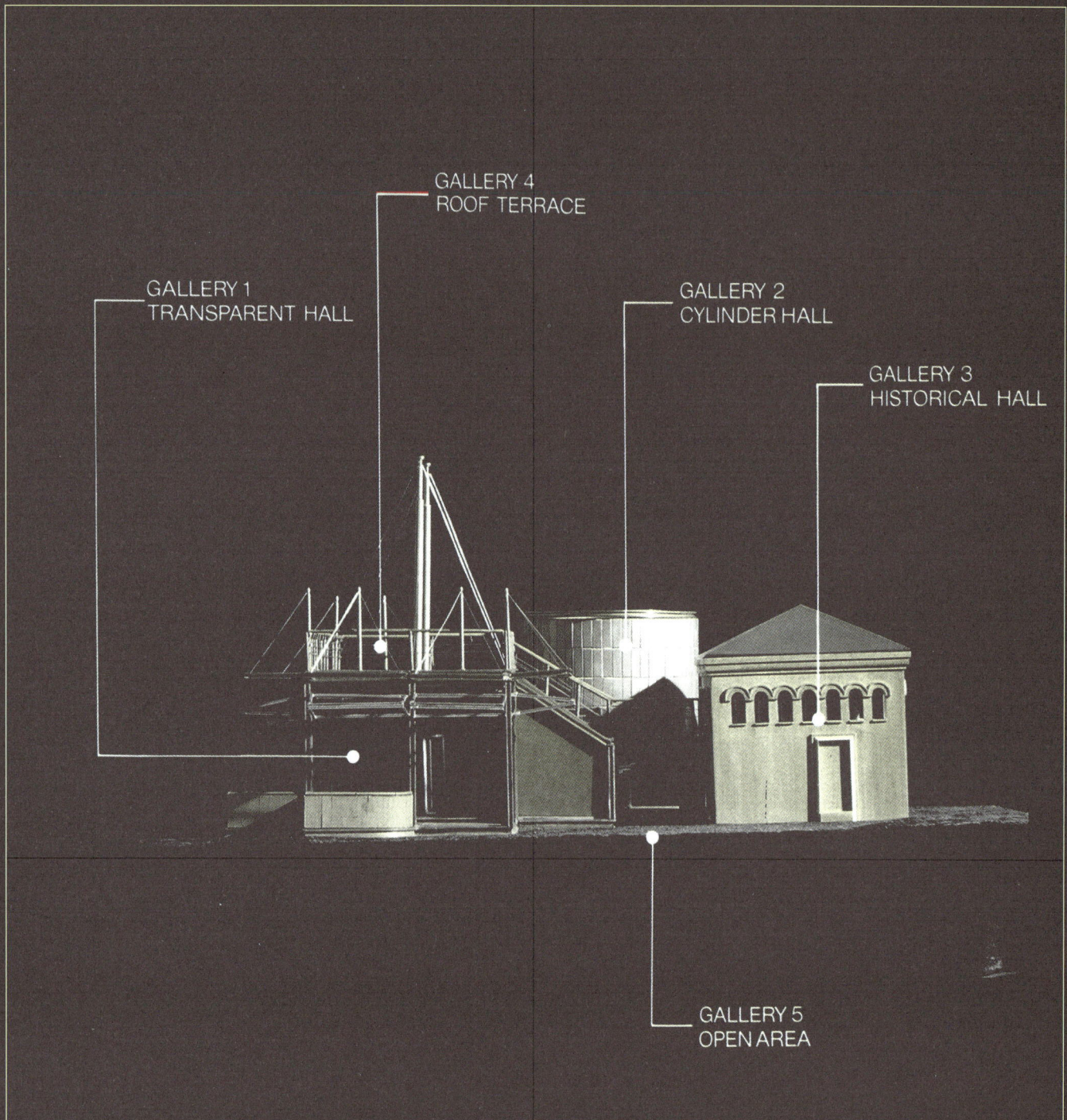

GALLERY 1 TRANSPARENT HALL
GALLERY 2 CYLINDER HALL
GALLERY 3 HISTORICAL HALL
GALLERY 4 ROOF TERRACE
GALLERY 5 OPEN AREA

By placing an abstract cylindrical structure parallel to the square-shaped old building, the design sought to merge the past and the future. A transparent grid was introduced to flow between these two volumes and the existing trees.

— Kim Seok Chul, "Korean Pavilion in Venice Biennale," *SPACE*, January 1995.

Ⓒ Kim Seok Chul. Courtesy of ARKO Arts Archive, Arts Council Korea (Contributor: Kim Seok Woo)

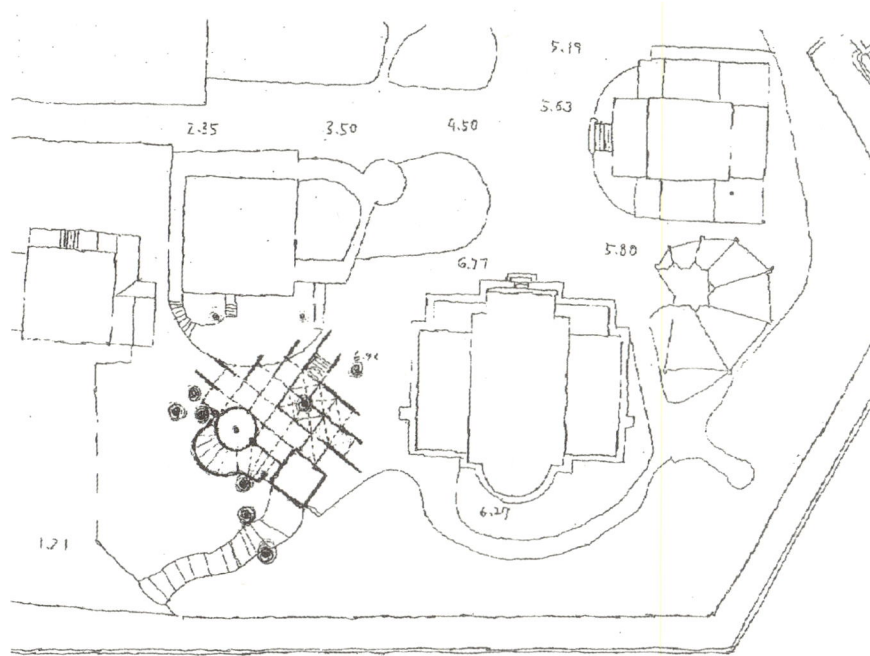

©Mancuso e Serena Architetti Associati. Courtesy of ARKO Arts Archive, Arts Council Korea.

EMBRACING THE HISTORY OF THE GIARDINI: Following the November 1993 agreement, the Korean Pavilion was approved for construction under the condition that the brick building would be restored in its original form. Kim Seok Chul and Franco Mancuso consciously responded to its historical significance. The sleek and transparent steel structure was not only a choice for openness but also a deliberate contrast to the opaque brick walls that bore the marks of time. Going further, to geometrically counterbalance the square form of the brick building, a cylindrical hall was proposed. Early designs reveal this juxtaposition of the square of history and the circle of the future, combined with a transparent grid. In the final proposal, the brick building was preserved as a historical element and was officially designated as the "Historical Hall."

The photographic image used in the front cover of the exhibition booklet in Spazio Olivetti.
©Kim Seok Chul. Courtesy of ARKO Arts Archive, Arts Council Korea
(Contributor: Kim Seok Woo)

The photographic image used in the back cover of the exhibition booklet in Spazio Olivetti.
©Kim Seok Chul. Courtesy of ARKO Arts Archive, Arts Council Korea
(Contributor: Kim Seok Woo)

Municipality of Venice's press release regarding the exhibition introducing the Korean Pavilion at Spazio Olivetti (November 1994)
ASAC, Fondo storico, Progetti speciali b. 5874 / Courtesy © Archivio Storico della Biennale di Venezia, ASAC

MUNICIPALITY OF VENICE
DEPARTMENT OF CULTURE
Press Office

...

SOUTH KOREA PAVILION
AT THE VENICE BIENNALE
project exhibition
Olivetti Space
November 9 - 20, 1994
9:30/12 - 15:30/18 (closed on Sundays)

...

organized by the Ministry of Culture and Sports of the Republic of Korea will be inaugurated and by the Municipality of Venice The exhibition, which will remain open to the public until November 20, 1994, with opening hours 9.30/12 - 15.30/18 (closed on Sundays), presents a series of sketches by the architect Kim Seok Chul for the new Korean pavilion.
  The area is identified at the top of the small artificial hill existing between the Japanese and German Pavilions, in a secluded position with respect to the large avenue that ends with the structure of Great Britain.
  The project proposes the recovery of the brick building, now very degraded, which will be incorporated into the new construction, thus composed of two intrinsically linked entities. ...

WHAT MATTERED TO THE CITIZENS OF VENICE: When the Korean Pavilion began construction and was exhibited at Spazio Olivetti, the first and foremost characteristic presented to Venetian citizens was neither publicness nor transparency, but rather the "recovery of the brick building" (il recupero dell'edificio in mattoni). This emphasis was further accentuated with the project booklet displayed at the exhibition. The front cover featured the new structure added to the brick building, while the back cover showcased the original brick structure itself, highlighting its preservation. These indicate that demonstrating respect for history was a strategic approach to gaining acceptance from the general Venetian public.

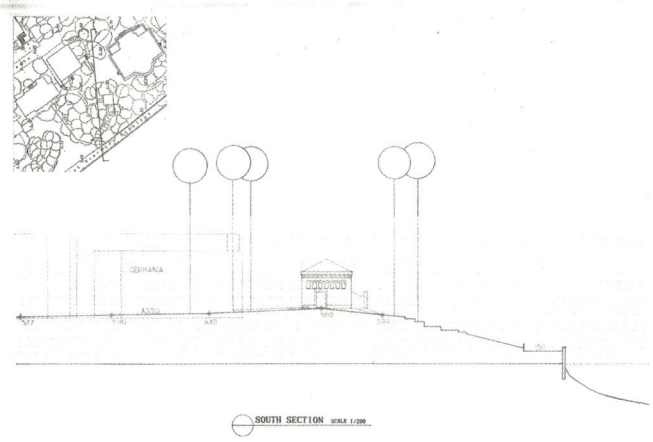

A Section showing the original brick building and the path leading to the embankment of San Marco Basin. ©Mancuso e Serena Architetti Associati. Courtesy of ARKO Arts Archive, Arts Council Korea.

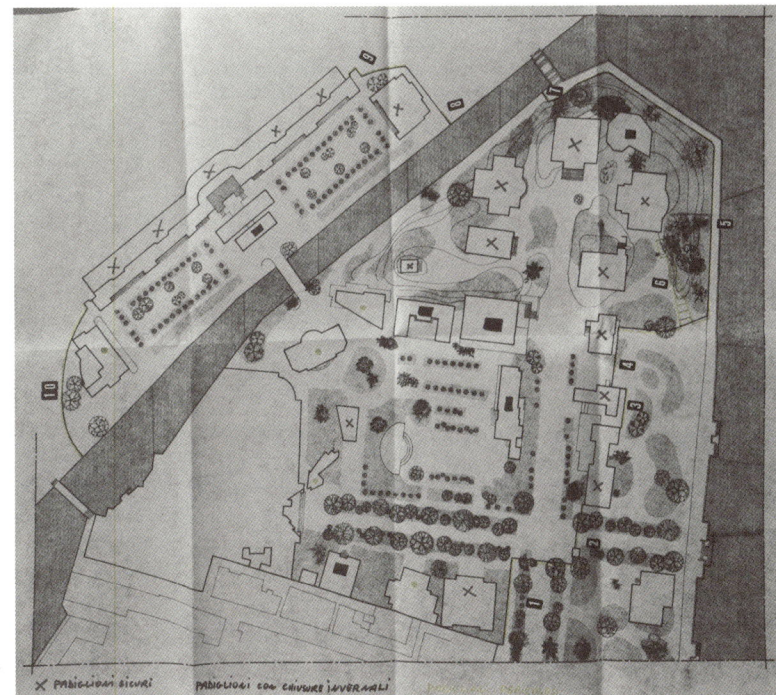

Safety evaluation of the buildings in Giardini (1981). The brick building, marked in yellow with the connected pathway, was categorized as "pericolosi"(dangerous). ASAC, Fondo storico, Lavori alle sedi, Padiglioni b. 9, Problema custodia Giardini / Courtesy © Archivio Storico della Biennale di Venezia, ASAC

A HOLE IN THE PARK: As the Giardini became increasingly enclosed with fences, accelerating its transformation into an enclave, the brick building caught the attention of the Biennale authorities. Their concerns centered on accessibility from the outside and the potential security risks it posed. A narrow path leading to the brick building had been present since at least 1910s, as seen in historical maps. This path remained intact even after fencing was introduced, meaning that by simply crossing one barrier, one could move from the Fondamenta along the San Marco Basin into the Giardini, directly reaching the brick building. The path's persistence may have been due to Giardini's unique constraints, where architectural interventions are particularly difficult to implement. For the Biennale authorities, this lingering access point was perceived as a vulnerability to the security of the Giardini. Notably, a 1981 document from the Biennale management explicitly depicts the brick building along with its connected narrow path, and categorizes them as "pericolosi" (dangerous).

The brick building and the pathway from Fondamenta before the construction of the Korean Pavilion
©Mancuso e Serena Architetti Associati. Courtesy of ARKO Arts Archive, Arts Council Korea.

©Mancuso e Serena Architetti Associati. Courtesy of ARKO Arts Archive, Arts Council Korea.

trees and vegetation.

The main access will be from the eastern side, by way of a planned loop in the contact point between the two volumes that flank the existing building. There will be an opening on the eastern side to allow access from the existing route from the embankment, and this is especially important should the diaphragm that today separates the Biennale from the Public Gardens be removed.

In this light, use of the pavilion on a continuous basis during the winter season has been taken into consideration. This led to two important distinguishing features. The first concerns the technical systems, as the pavilion is to be fitted with an efficient winter heating system, in addition to summer air conditioning. The second concerns the perimeter closure system. Unlike the other pavilions which are usually protected during the winter by temporary closure systems, the Korean pavilion will be equipped with an external protection system which is an integral part of the building, and which makes opening and closing operations fast and efficient. The system is based on external folding wooden elements. When open they are hung on external cornices, as in traditional Korean architecture.

The building's layout is simple, and the exhibition area runs in a longitudinal sense (east-west), emphasising the visual projection towards St. Mark's Basin. This space and the existing building are connected by a cylindrical body, light, transparent and taller, as it includes the technical systems

©Mancuso e Serena Architetti Associati. Courtesy of ARKO Arts Archive, Arts Council Korea.

FROM A HOLE IN THE PARK TO AN ACCESS POINT FOR REOPENING THE GIARDINI: Conversely, the hole in the park could also be seen as a clue for reintroducing public openness. Beyond merely integrating the Korean Pavilion into its physical surroundings, the architects saw an opportunity to shape it as an open public space that would enhance Giardini's accessibility on a programmatic level. The narrow path and the brick building became key references in the circulation design. By utilizing the existing pathway leading outward, the Korean Pavilion was envisioned as an access point that could return a part of the Giardini to Venetian citizens, especially during non-Biennale periods.

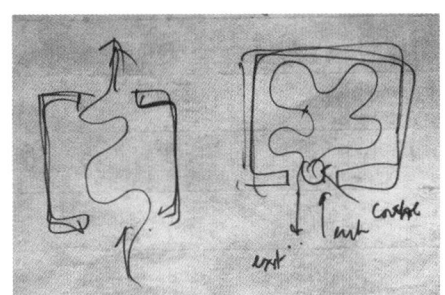

TWO ENTRANCES CUTTING THROUGH A TRANSPARENT STRUCTURE: Today, the Korean Pavilion is perceived as having a single entrance — the main entrance that exhibition visitors encounter when entering from the inside of Giardini, between the Japanese and German Pavilions. However, during the design phase, the pavilion was conceived as a permanently accessible public space that could be directly reached via the narrow path. This meant that its face toward Fondamenta, the façade seen by Venetian citizens approaching from outside was just as significant. The Korean Pavilion was designed as a transparent structure "on the street," bridging the inside and outside of the park. Accordingly, two entrances of equal importance were designated on either side of the path that cut through the structure. Indeed, design archives of the pavilion contain more perspective sketches from the narrow path than from within Giardini, underscoring the importance placed on its accessibility from the city outside.

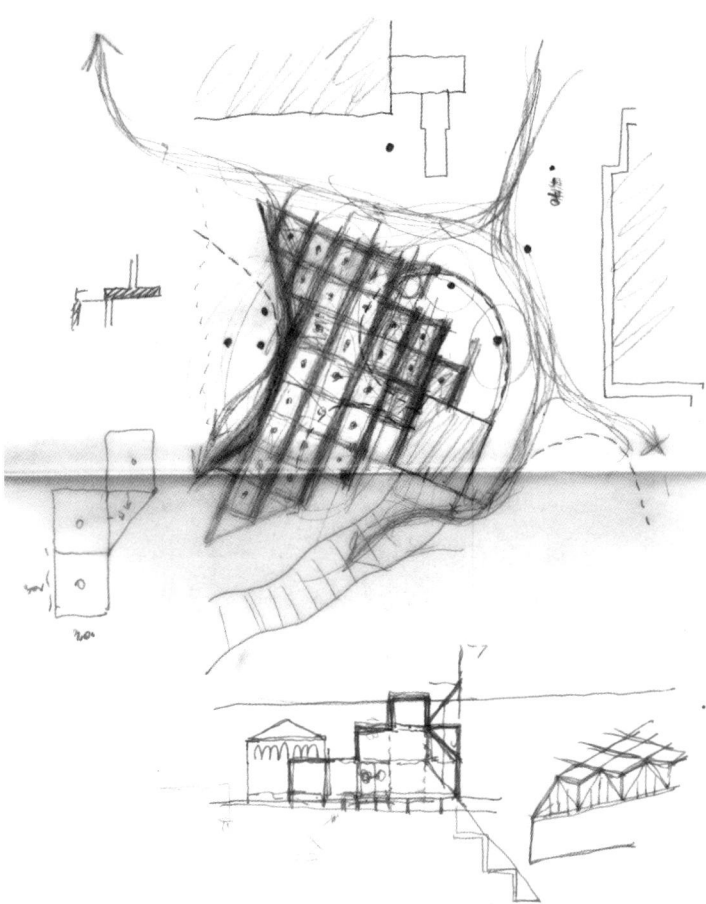

©Mancuso e Serena Architetti Associati. Courtesy of ARKO Arts Archive, Arts Council Korea.

A perspective sketch, approaching from outside of the Giardini via existing pathway ©Mancuso e Serena Architetti Associati. Courtesy of ARKO Arts Archive, Arts Council Korea.

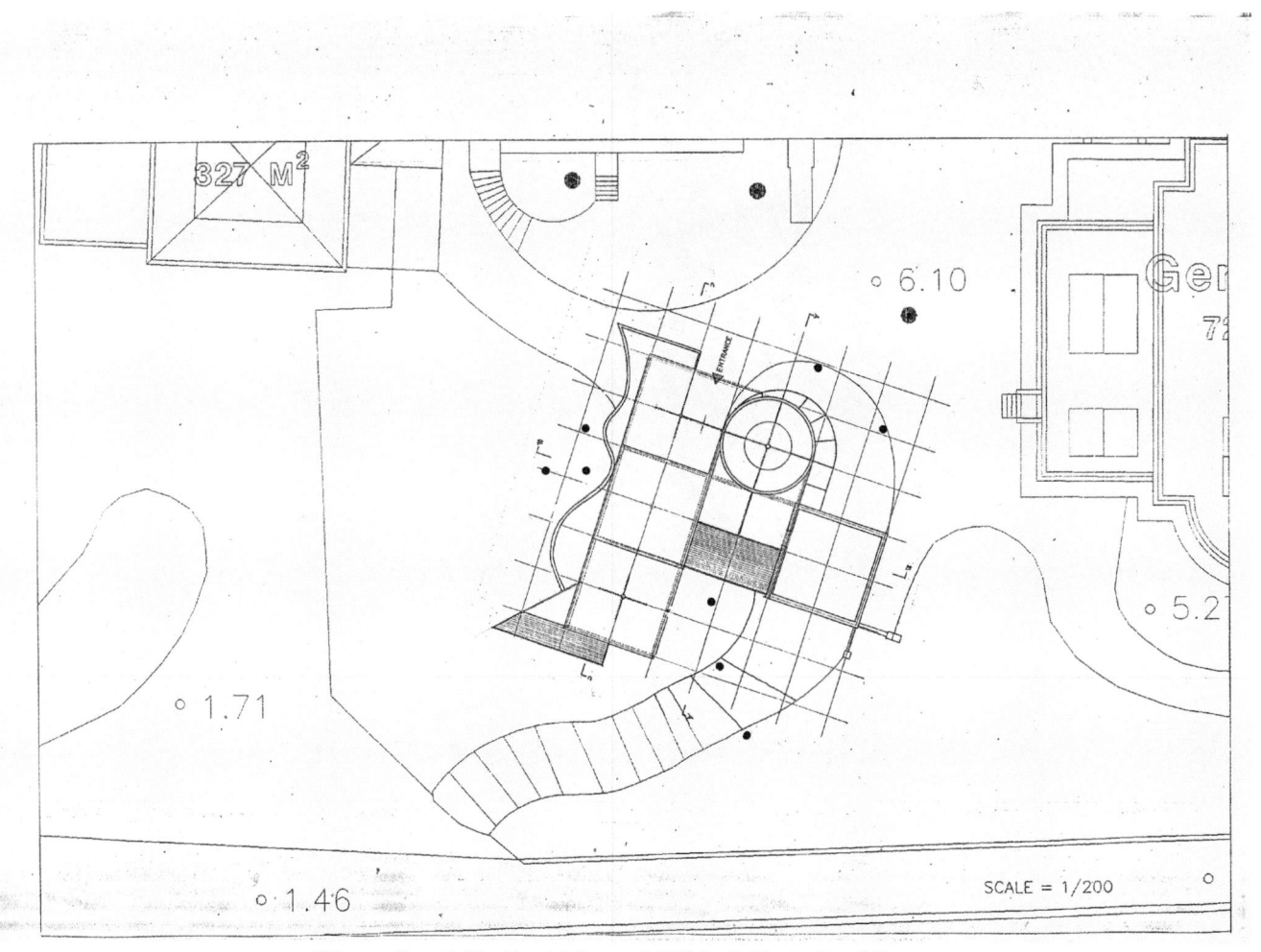

©Mancuso e Serena Architetti Associati. Courtesy of ARKO Arts Archive, Arts Council Korea.

SHAPING THE FLOW: The awareness of the narrow path left traces not only in the circulation plan but also in the very form of the building. While the cylindrical mass was primarily a geometric counterbalance to the square-shaped brick building, the dual-cylinder layout also served another purpose. Its planar configuration absorbed the winding trajectory of the narrow path, transforming its organic flow into the spatial composition of the building itself.

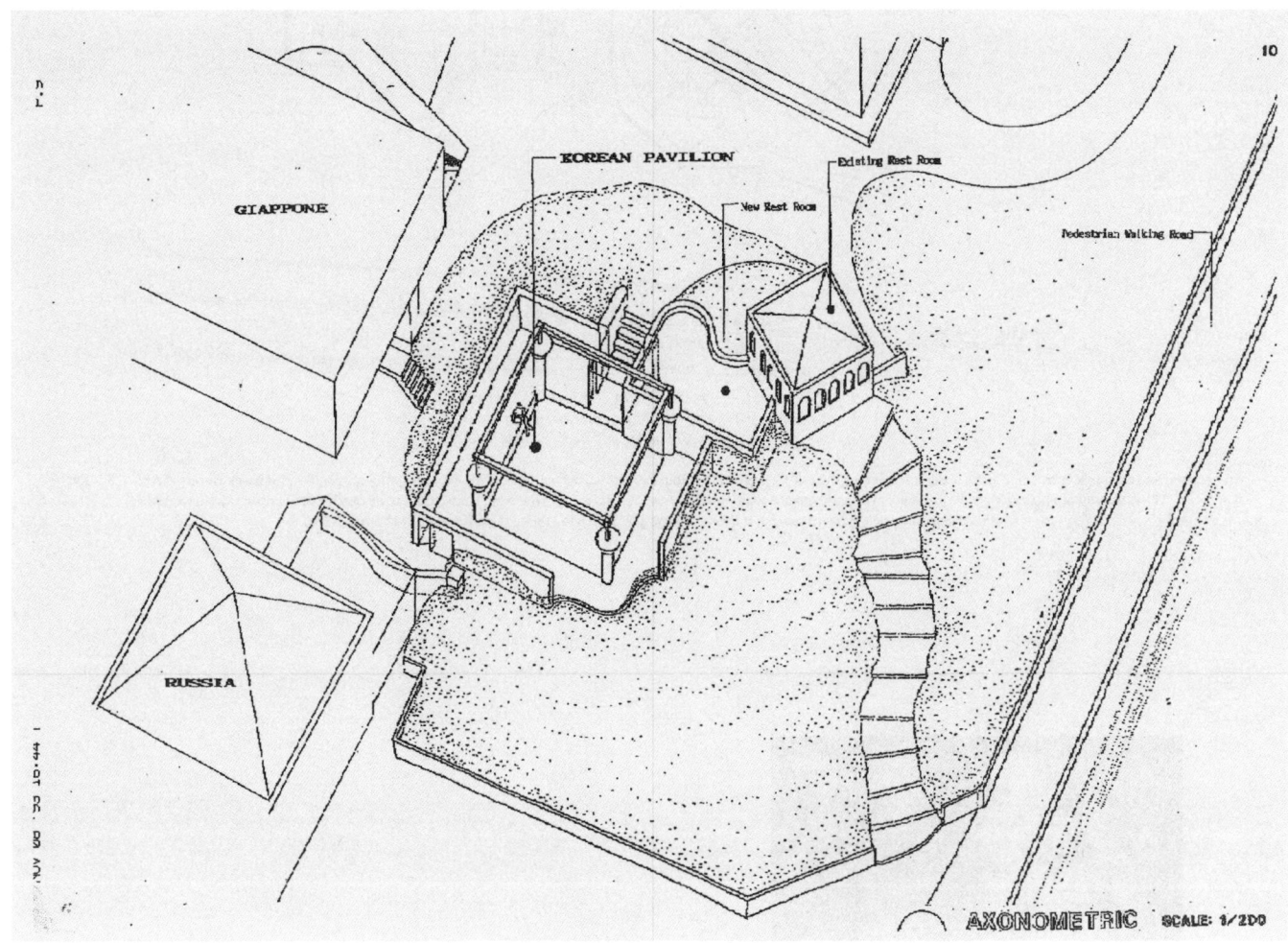

The underground proposal for the Korean Pavilion by Kim Seok Chul,
sent to Franco Mancuso on November 9th, 1993.
©Mancuso e Serena Architetti Associati. Courtesy of ARKO Arts Archive, Arts Council Korea.

The awareness of the narrow path appears from the very early
stages of the project. In the initial underground pavilion
design, little attention was given to entry from within
the Giardini. Instead, it appears that the narrow path was
envisioned as the main access route, guiding citizens through
the brick building and into the pavilion. At the point where
the path ascends, only four columns stand in an open and
transparent void. A solitary figure sitting in this empty space
underscores the pavilion's true identity — This structure was
conceived first as a public space, even before being an art
gallery.

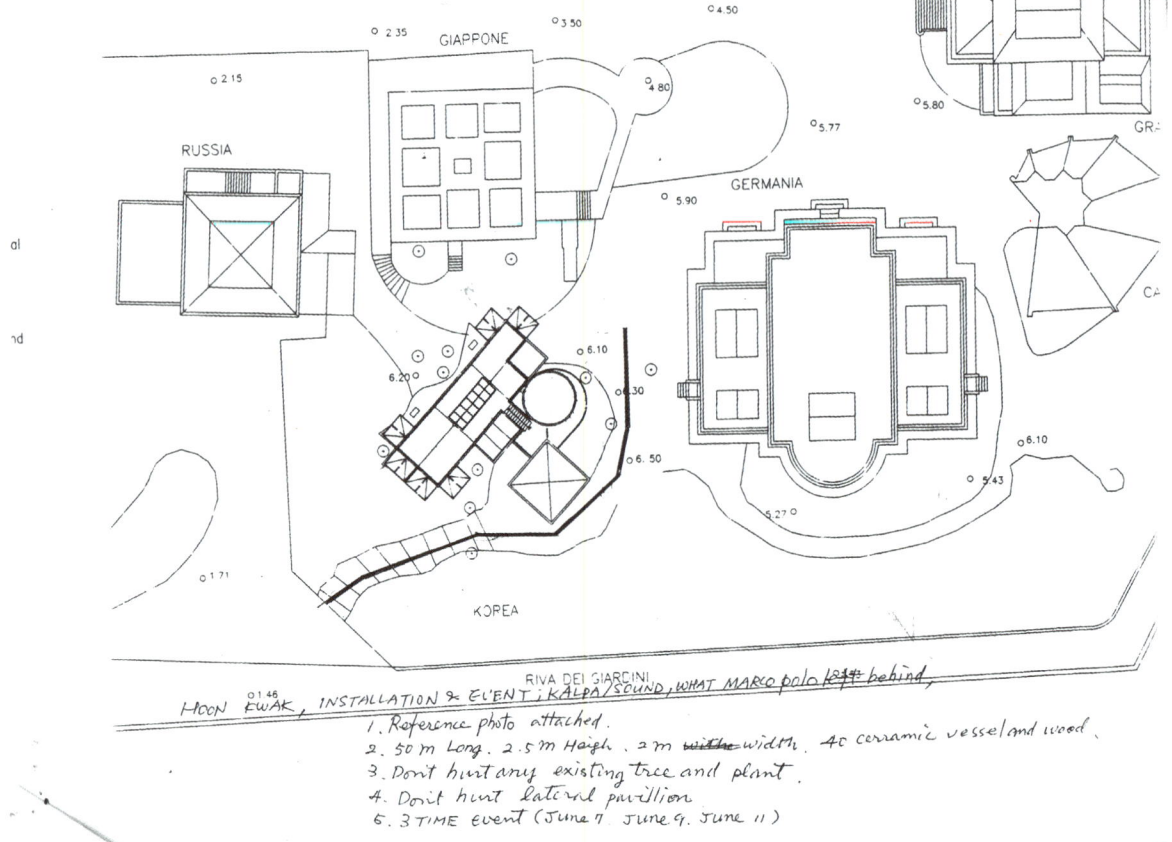

An excerpt from a letter sent by Lee Yil, the inaugural commissioner of the Korean Pavilion, to neighboring national pavilions, requesting cooperation for Kwak Hoon's 'KAPLA/SOUND: What Marco Polo Left Behind' performance. The performance route was suggested to pass the narrow path, connecting to the outside of Giardini (May 31st, 1995)
ASAC, Fondo storico, Progetti speciali b. 5874 / Courtesy © Archivio Storico della Biennale di Venezia, ASAC

Kwak Hoon's 'KAPLA/SOUND: What Marco Polo Left Behind' performance (1995)
©Mancuso e Serena Architetti Associati. Courtesy of ARKO Arts Archive, Arts Council Korea.

A FORGOTTEN VISION: With the opening of the Korean Pavilion in 1995, the International Art Exhibition was held simultaneously. Four artists were invited, with three occupying distinct sections of the pavilion: the Historical Hall (brick building), the Cylinder Hall, and the Transparent Hall (the steel grid structure). However, the work of the fourth artist, Kwak Hoon, was not exhibited in the Rooftop Hall, which corresponded the "deck" of the pavilion and was intended as its fourth hall. Instead, his work was placed along the narrow path. 'KAPLA/SOUND: What Marco Polo Left Behind' began at the front of the Korean Pavilion, tracing the narrow path and leading toward the bank of the San Marco Basin, the very waters Marco Polo once crossed on his journey to the East. If Nam June Paik's 'Marco Polo,' installed in front of the brick building two years prior, had reterritorialized the Korean Pavilion's site as a land of the future, Marco Polo conjured by Kwak Hoon embodied and materialized this vision. Through the new architectural device named the Korean Pavilion, this hill would extend beyond the Giardini, and further beyond Venice itself.

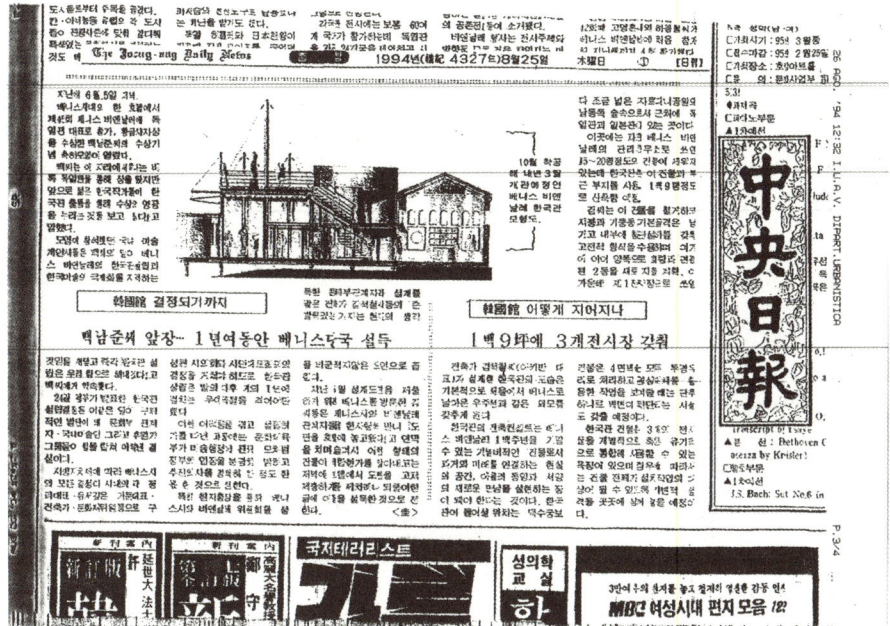

©Mancuso e Serena Architetti Associati. Courtesy of ARKO Arts Archive, Arts Council Korea.

The Korean Pavilion, designed by architect Kim Seok Chul (principal of ARCHIBAN), essentially takes on the appearance of a spaceship that has flown from Seoul to Venice.
The architectural concept of the pavilion is to serve as a monumental structure commemorating the 100th anniversary of the Venice Biennale. It is envisioned as a space that connects the past and the future, while also becoming a site for a new encounter between East and West.

The Joongang Daily, August 25th, 1994.

©Mancuso e Serena Architetti Associati. Courtesy of ARKO Arts Archive, Arts Council Korea.

The project involves incorporating a small brick building from the 1930s into an artificial hill that will house public services. The new structure will be airy, transparent, and lightweight, supported on stilts.

…

It will be dismantlable and environmentally friendly, without harming existing trees. The technology for the installations will be provided by Korean industry, ensuring air conditioning that will allow year-round use of the space.

This initiative is closely tied to local government efforts to revitalize the Giardini and open Italy's Museum of Contemporary Art pavilion. The mayor enthusiastically welcomed Korea's initiative but emphasized that it is not just about culture: "Venice has always been oriented towards collaboration with the East," said Massimo Cacciari. "It has enormous potential for developing cultural and artistic relationships as well as economic and social ties with these countries."

"Korea is becoming one of the world's leading economic powers, continued the mayor, and soon it will find itself in a situation similar to that of post-war Japan: investing abroad to distribute its profits. In Venice, especially in Marghera, all conditions exist to concentrate new initiatives."

The message South Korea intends to launch in Europe with its Biennale pavilion could also have political significance: "Our country is the only one in the world divided into two," Minister Lin Min Sup reminded everyone. "We hope to increase cultural and economic exchanges with North Korea." And this pavilion must also contribute to reconciliation.

Il Gazzettino, November 9th, 1994.

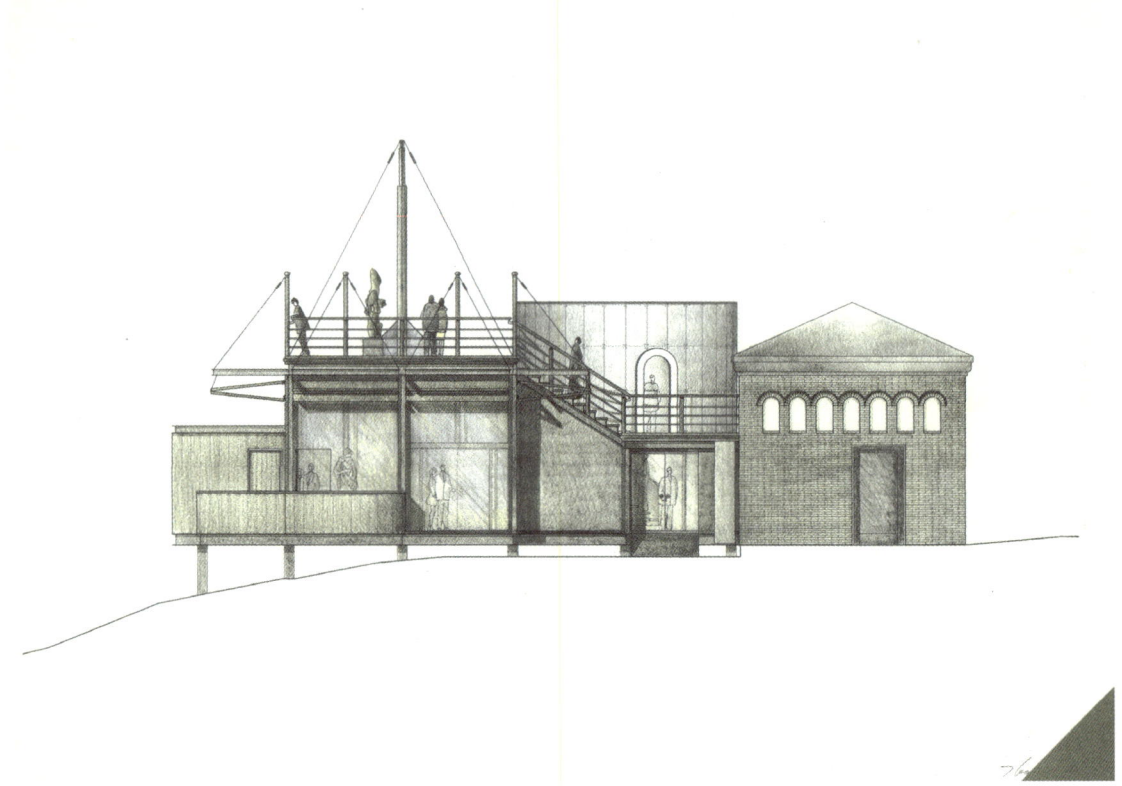

Final stage elevation design of the Korean Pavilion
©Kim Seok Woo. Courtesy of ARKO Arts Archive, Arts Council Korea.

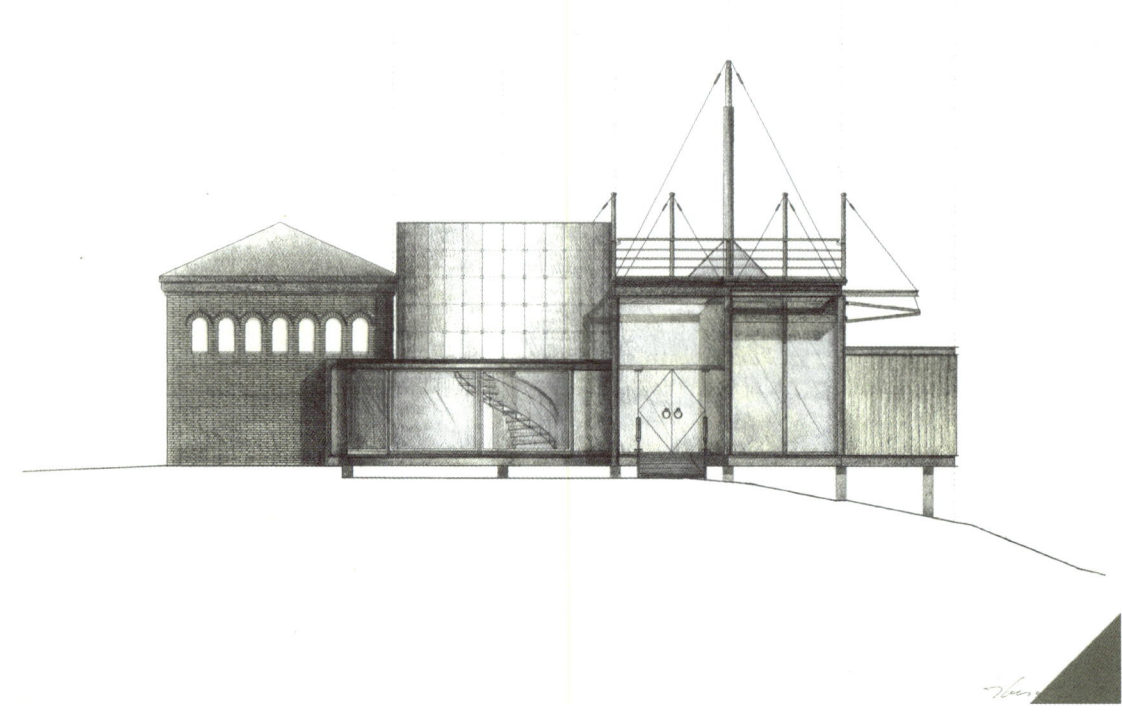

```
JUL 05 '96 09:49 T                                              P.6
```

> Opere e progetti
> **SEOK CHUL KIM AND ASSOCIATES con FRANCO MANCUSO**
>
> Committente
> **THE MINISTRY OF CULTURE AND SPORTS OF KOREA**
> **THE KOREAN CULTURE AND ARTS FOUNDATION**
>
> Progetto
> **SEOK CHUL KIM AND ASSOCIATES**
> **Arch. Seok Chul Kim (Seoul)**
> **Arch. Franco Mancuso (Venezia)**
>
> Direzione Lavori
> **Arch. Franco Mancuso**
>
> Coordinamento della progettazione
> **Rinio Bruttomesso, Kyung Soo Kim**
>
> Progettazione esecutiva
> **Chang Geun Park, Jin Young Choi, Seok Woo Kim**
>
> Realizzazione delle strutture
> **Chang Nam Lee**
>
> Impresa construttrice
> **ICCEM**

△ CONTENTS OF CREDIT TO BE RELIEVED ON COPPER PLATE.

PAGE 6 OF 13

```
05-LUG-1996 02:45          9              95%              P.06
```

Unrealized copper nameplate of the Korean Pavilion, planned in 1996.
©Mancuso e Serena Architetti Associati. Courtesy of ARKO Arts Archive, Arts Council Korea.

1995 →

2025 →

AN INHERENT FUTURE:
FOR PAVILION(S) TO EMBRACE THE WORLD

©Mancuso e Serena Architetti Associati. Courtesy of ARKO Arts Archive, Arts Council Korea.

> Who's this moving alive over the moor?
>
> An old man seeking and finding a difficulty.
>
> ...
>
> tussocks, minute flies,
>   wind, wings, roots
>
> He consults his map. A huge rain-coloured wilderness.
> This must be the stones, the sudden movement,
> the sound of frogs singing in the new year.
> Who's this issuing from the earth?
>
> — Alice Oswald, *Dart* (2002)

Frogs emerge in cultures across the world as symbols of life and fertility, fortune, resurrection, and magic.

...

In Egyptian mythology, the Frog Goddess Heqet is a creator of life as the force that brings life into the womb, mirroring the regenerative life of the Nile as it blew past its banks each spring, bringing with it not only water and rich soil, but thousands of spring frogs.

...

Frogs are creatures of metamorphosis, beginning from egg to tadpole with tail and gills, and then on to frog with legs and lungs as they move from their watery womb birthplace onto land.

What an extraordinary process to muse on, and to break open our imaginings of how we may shift in our lifetimes, finding fluid fluency in our shifting understandings of ourselves and the world around us. And in winter frogs hibernate, but it is no simple sleep of a tucked-in bear dreaming of summer berries. Instead, frogs freeze solid as rocks—a trick of glucose and ice crystals forming outside cells—and when they thaw on an early warm spring evening, make their way across the forest floor and find a pond or vernal pool. Literally and metaphorically, frogs rise from what looks and feels like death to sing spring into being.

— Leigh Melander, "Life, Resurrection, and the Mythic Teachings of Frogs"

Humanity's situation with respect to climate change is sometimes compared to that of a frog in a slowly boiling pot of water, meaning that change will happen too gradually for us to appreciate the likelihood of catastrophe and act before it is too late.

— Simon Sharpe, "Telling the Boiling Frog What He Needs to Know: Why Climate Change Risks Should be Plotted as Probability Over Time," *Geoscience Communication*, May 29, 2019.

```
EMBRACING A SHARED WORLD: The last pavilion built in the
Giardini, born from the optimism that the borders of the world
and the fences of the Biennale could be dismantled, has now
turned thirty. Over the past three decades, the frontlines
of the world have shifted in both location and nature. The
aftermath of the Cold War has fully dissipated, giving way
to a struggle with the "boiling" world. The optimism of a
new civilization has vanished like smoke. Frankenstein-like
artificial crises disrupt the natural world and widen social
inequalities to extremes. As social safety nets collapse and
survival itself becomes precarious, fascism and dictatorship
resurface, and the imagination of "the other" dries up. The
existential rationale of independent national pavilions as
public entities seems more fragile than ever. However, just as
the future does not lie in blind optimism, it does not exist
in a total reset.
    Instead, let us begin by embracing what has already
arrived. We may find new possibilities by revisiting and
reinterpreting the Korean Pavilion's longstanding approach to
engaging with the shared elements — land, trees, sea, and sky
— while layering strata of imagination upon them. Through this
act of "unbuilding" a pavilion imbued with inherent futures,
we may catch a glimpse of path for pavilion(s) to embrace a
shared world.
```

Lee Dammy's installation, which begins at the entrance of the Korean Pavilion, holds symbolic significance as a loop that circulates the beginning and end of *Little Toad, Little Toad*. 'Overwriting, Overriding' brings forth the hidden entities that have silently coexisted with the pavilion, offering an opportunity to look back on the pavilion's history. Among the four narrators selected by Dammy are the honey locust tree guarding the Korean Pavilion, and Mucca the cat who roams the pavilion as if it were its own home. 'Overwriting, Overriding' reveals that the pavilion has been home to broader entities; not only to the architects, curators, artists, or artworks. Through this speculative approach, Lee Dammy re-examines and "overwrites" the Korean pavilion, inserting hole-like fractures into its established narrative, inviting the audience to step into Alice's rabbit hole of this exhibition.

— Chung Dahyoung, Curatorial Essay (2024)

© Lee Dammy

```
p.c. AL COMANDO VV.UU. - SEDE
    ALL' ASSESSORATO PATRIMONIO - SEDE
    ALL' ASSESSORATO LL.PP.  COMUNE DI VENEZIA
                             ASSESSORATO EDILIZIA PRIVATA
```

```
                    AUTORIZZAZIONE

         Interventi ai sensi dell'art. 2 del Regolamento Edilizio
         Commi b) c) d) e) f) g) della Legge 5.8.78 N.457 (art.48)
                    e L.R. 27.6.85 N.61 (art.76)

PROT. N.   94/6865                    Prot. generale  94/81191
                                      Prot. Legge 373

              IL DIRIGENTE DEL SERVIZIO

Vista la domanda della Ditta in data 14/06/94
Visto il parere della Comm. Salvaguardia Venezia N. 79/17572    del 08/08/94

              AUTORIZZA  LA  DITTA

REPUBBLICA COREA DEL SUD  COMMITTENTE
WOONG HO LEE-MINISTRO DELLA CULTURA  COMMITTENTE


ad eseguire in CASTELLO (VENEZIA)
        sez. VENEZIA         fg.         mapp.

i seguenti lavori:

COSTRUZIONE PADIGLIONE PROVVISORIO PRESSO I GIARDINI DELLA BIENNALE DI VENE_
ZIA, COME DA PROGETTO E RELAZIONE TECNICA A FIRMA DEGLI ARCHITETTTI SEOK CHUL
KIM E FRANCO MANCUSO.
```

```
           MUNICIPALITY OF VENICE
           PRIVATE BUILDING DEPARTMENT
...

AUTHORIZES THE FIRM
    REPUBLIC OF SOUTH KOREA       CLIENT
    LEE WOONG HO - MINISTER OF CULTURE    CLIENT


to carry out in CASTELLO (VENICE)

CONSTRUCTION OF A TEMPORARY PAVILION AT THE VENICE BIENNALE
GARDENS, AS PER THE PROJECT AND TECHNICAL REPORT SIGNED BY
ARCHITECTS KIM SEOK CHUL AND FRANCO MANCUSO.

Conditions
    — THE PAVILION IS DISMANTLED
    — NO TREES ARE TO BE FELL DOWN AND THE ALTIMETRIC
    CONTOUR OF THE LAND IS NOT TO BE CHANGED
    — THE PAVILION IS TO BE DISMANTLED BY 12/31/1998,
    PROVIDING FOR THE RESTORATION OF THE LAND TO ITS
    ORIGINAL CONDITIONS, IF, BY THAT DATE, A SPECIFIC
    LANDSCAPING INSTRUMENT IS NOT APPROVED WHICH CONFIRMS
    ITS PERMANENCE.
```

Final permission of the construction of the Korean Pavilion, published by Private Building Department of Municipality of Venice, following the approval of Venice Safeguard Commission.
©Mancuso e Serena Architetti Associati. Courtesy of ARKO Arts Archive, Arts Council Korea.

```
A VANISHING STRUCTURE THAT TOUCHES NEITHER LAND NOR TREES:
As evidenced by the construction permit issued by the Venice
Safeguard Commission and the Private Building Department,
three conditions were ultimately attached to the establishment
of the Korean Pavilion:

1.  No trees were to be disturbed.
2.  The terrain was not to be altered.
3.  The pavilion had to be built as a "temporary structure"
    that could be dismantled at any time.

Even by today's standards of sustainability in architecture —
a concept barely discussed at the time — these were extreme
requirements.
```

Venetians inserting the wooden pile on the swampy terrain. Drawn by G. Grevembroch in mid 18th Century

ARTIFICIALITY MADE OF NATURE: The seemingly excessive condition of constructing a building in the middle of a forest without disturbing a single tree or altering the land stems from Venice's unique circumstances. Venice itself is an artificial land created out of nature. Located in a lagoon, the city was built by early settlers who fled to the marshlands and sought to make them livable. Their first task was to establish an artificial foundation. To do so, they transported thousands of oak trees from the mainland region of Friuli, driving them deep into the soft ground as piles, upon which they meticulously stacked bricks to form solid structures.

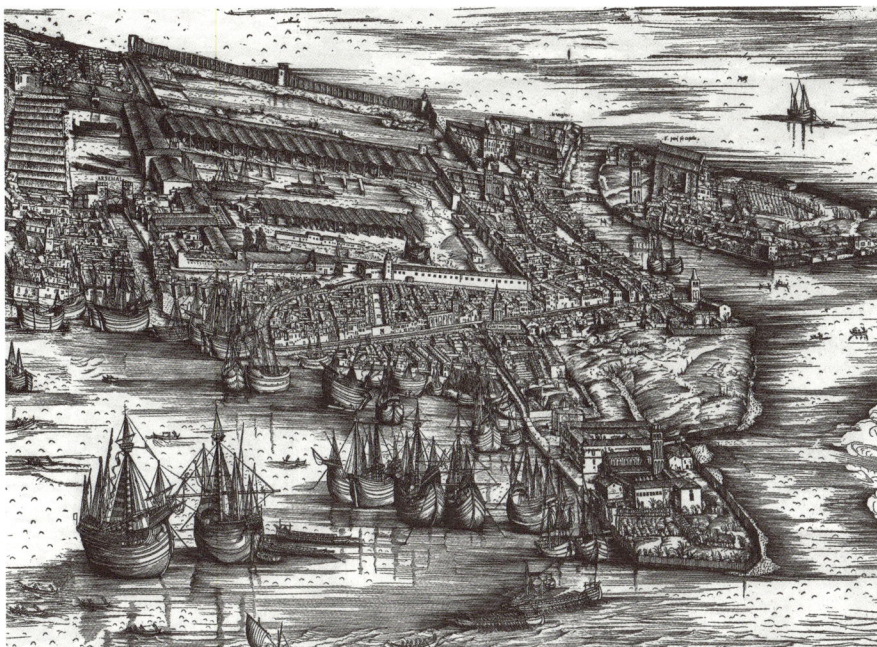

Jacopo de' Barbari, 'Veduta di Venezia a volo d'uccello' (1498)

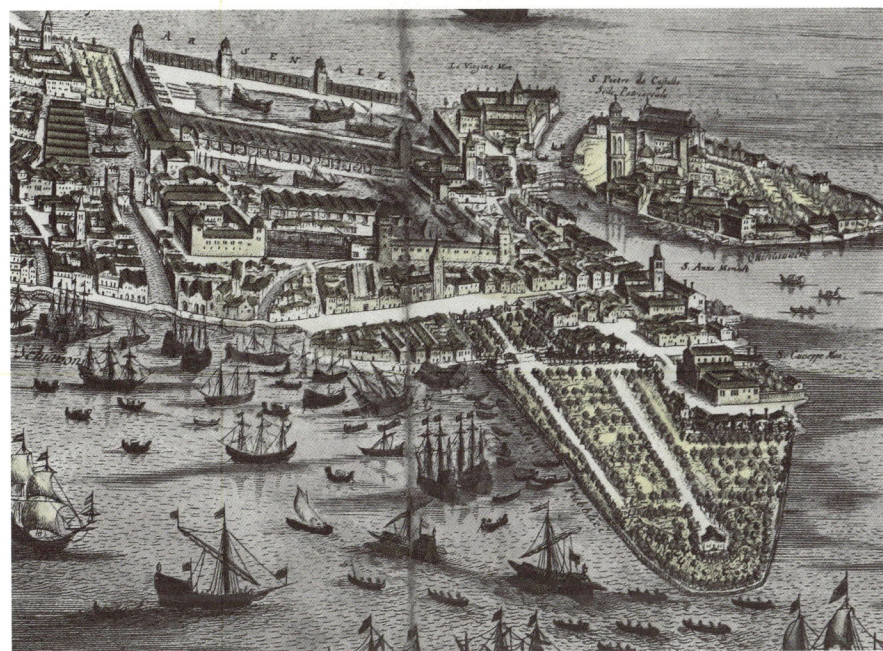

Marco Sebastiano Giampiccoli, 'Veduta di Venezia' (1810)

non arat, non seminat, non vendemiat (No cultivation, no sowing, no harvesting)

– Fernand Braudel, *The Material Civilization and Capitalism III* (1997)

The Venetians are jealous of every tree.

– Enrico Peressutti of BBPR, architect of the Canadian Pavilion, in a letter to the National Gallery of Canada.

ARTIFICIALLY CREATED NATURE: For this reason, natural green spaces rarely formed spontaneously in Venice. This is what makes the greenery of Giardini so valuable. In essence, the Giardini is a nature that was artificially created — living trees planted upon land built from dead trees. For Venetians, who lived on barren ground where "neither cultivation, nor sowing, nor harvesting" was possible, the trees densely planted by Napoleon in Giardini became an irreplaceable resource, something they could not easily afford to sacrifice.

Project Description

Traditionally, the Korean thought of beauty always got along with nature. This faith often seems to be lost because of the fast economic development. therefore koreans hesitate about the harmony between the building and the nature. Following this, the project begins from not destroying the existing site.

The site is located in the backyard of the main street. It looks like a deserted area. However, we find great possibilities for it to become a new attractive scenery. The quietness of the site can be converted into the merit that korean traditional architecture has in common.

The basic concept of the structure is transparency. This transparency supports the frequent communication between the outside and the inside. The basic idea about how the pavilion is used, comes from how the korean traditional festival has been held. Korean people gathered every five or six days in which the gathering became the natural market. The pavilion is located on the street so that in common days it is used as a secret and quiet memorial-like building and in special days it will be the strong supporter for maximizing the happiness and surprise.

Proposal:

This is a transparent structure lowered two meters into the ground, with respects to the sea level. You enter the building from the level between the Japanese and the german pavilions. Once inside the pavilion, you are 2 meters under the level of the entrance.
The pavilion has 4 columns. Transparent glass roofs that slide, in the form of an arch are installed when there is a festival, to establish a temporary exhibition hall. When there is no festival, the roof is taken down. The walls are permanent structures, but not visible from the level of the entrance. The walls are partly transparent glass structures with an overhang, which comes up above the ground level.
When there is no festival, the pavilion is a rear garden for people to enjoy, walk through and rest. A glass window with a view towards the lagoon beautifies the garden structure.
This pavilion is a circulating garden in organic form that is transparent from the ground level of the entrance.
The existing restroom on the site is restored for comfortable use. A part of the new pavilion is used as an additional space for the showers.

Traditionally, the Korean thought of beauty always got along with nature. This faith often seems to be lost because of the fast economic development. Therefore Koreans hesitate about the harmony between the building and the nature. Following this, the project begins from not destroying the existing site.

Proposal for the Korean Pavilion by Kim Seok Chul, sent to Franco Mancuso on November 2nd, 1993. ©Mancuso e Serena Architetti Associati. Courtesy of ARKO Arts Archive, Arts Council Korea.

A PROACTIVE INTERPRETATION OF CONSTRAINTS: The requirement to preserve the terrain and avoid cutting down even a single tree might seem nearly impossible even by today's standards. However, within the Korean Pavilion archives, complaints from the architects are scarce. Instead, they embraced respect for nature as a core architectural identity. As previously explored, "transparency" — blending into the vegetation of the park and the waters of the San Marco Basin — was already a fundamental concept of the design. Respect for nature was even regarded as an additional concept that could further strengthen the expression of Korean-ness.

① The different levels of the site

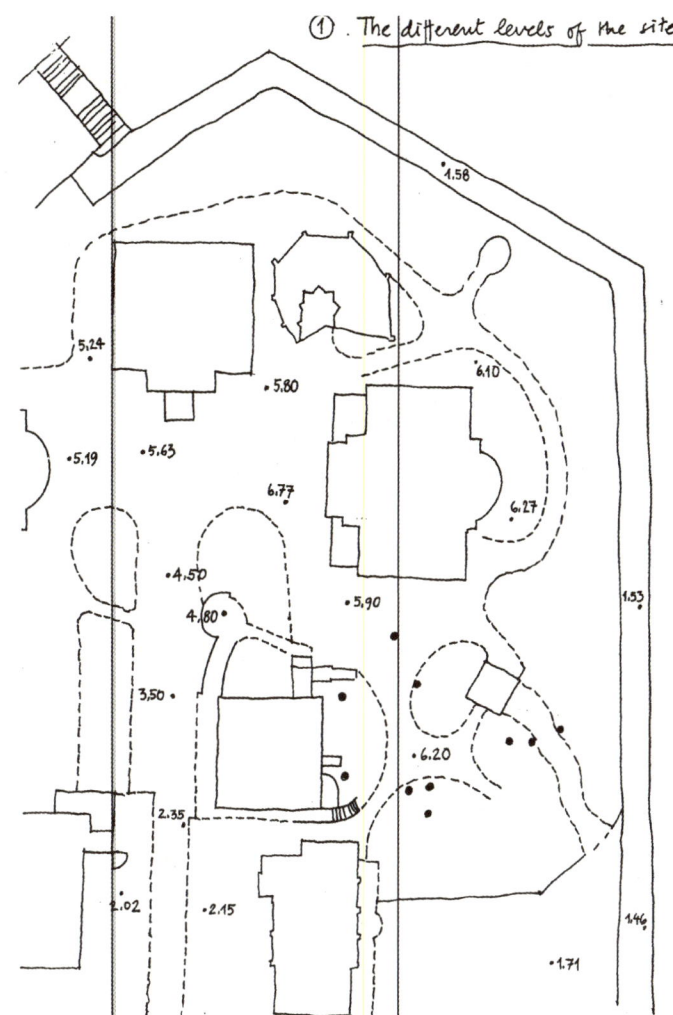

©Mancuso e Serena Architetti Associati. Courtesy of ARKO Arts Archive, Arts Council Korea.

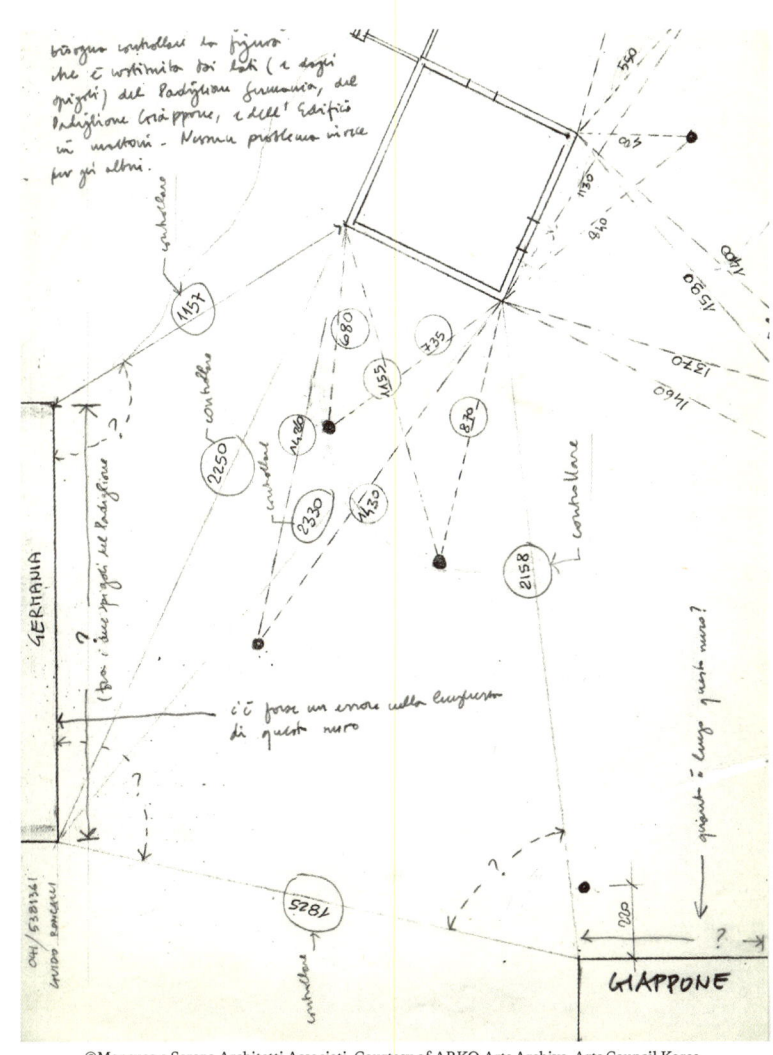

©Mancuso e Serena Architetti Associati. Courtesy of ARKO Arts Archive, Arts Council Korea.

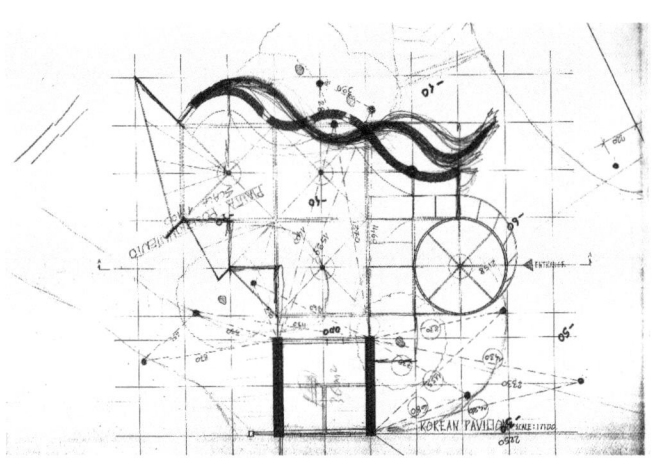 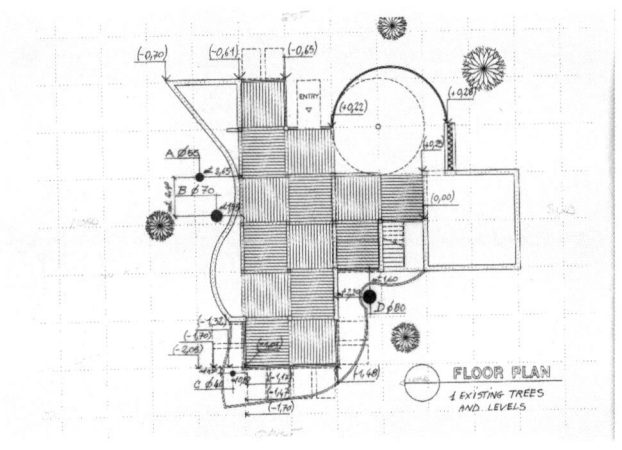

ARCHITECTURE WHERE THE PROCESS OF RESPECTING TREES AND LAND BECOMES THE DESIGN ITSELF: One of the first requests Kim Seok Chul made after inviting Franco Mancuso as a co-architect was a detailed survey of the site's elevation and the precise location of trees. The act of identifying and engaging with the trees and land was not merely a preliminary step but an ongoing process that continued from the inception of the pavilion to its completion, ultimately elevating itself into the very essence of the design.

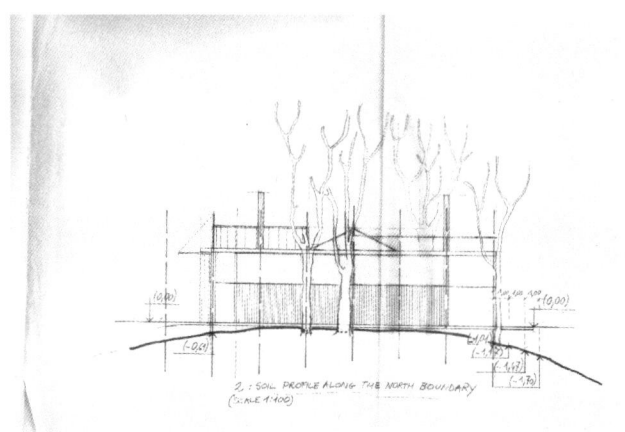 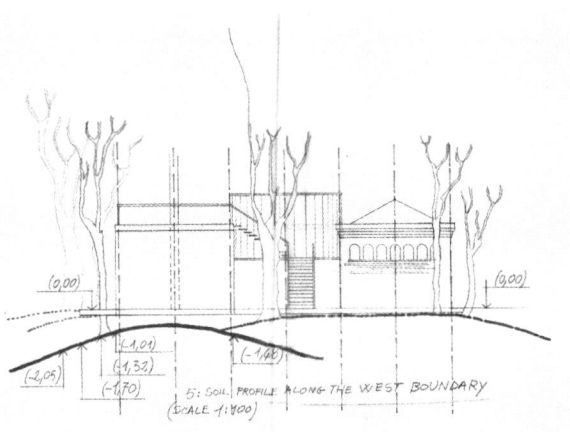

READING THE LAND AS IT IS: Flattening and erasing the natural characteristics and contour of a site is often the first step in architectural construction. However, the Korean Pavilion had to be built without altering the natural contours of the land. As a result, accurately identifying and understanding the existing elevation points remained a recurring challenge throughout the design process.

©Mancuso e Serena Architetti Associati. Courtesy of ARKO Arts Archive, Arts Council Korea.

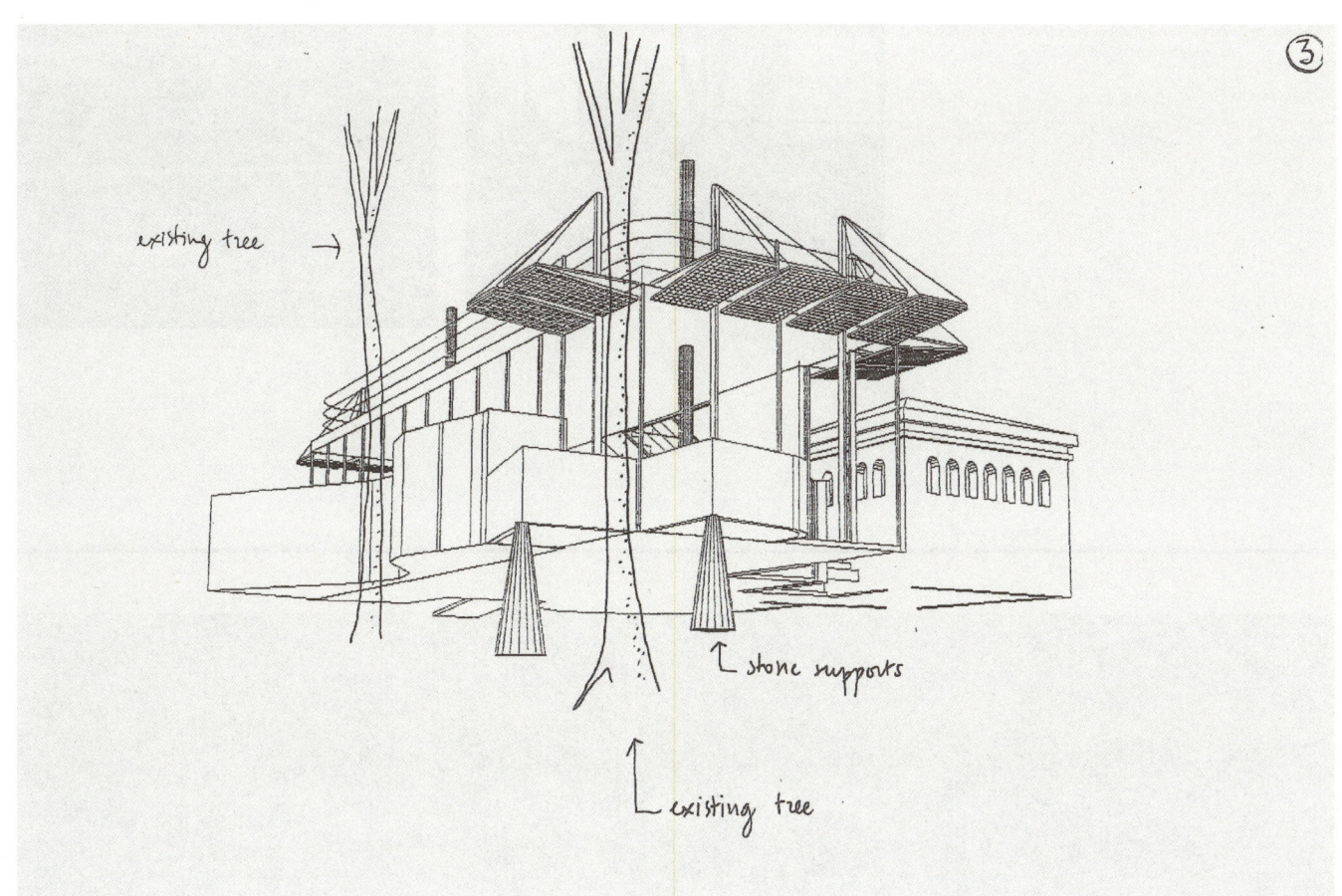

©Mancuso e Serena Architetti Associati. Courtesy of ARKO Arts Archive, Arts Council Korea.

©Mancuso e Serena Architetti Associati. Courtesy of ARKO Arts Archive, Arts Council Korea.

2024-10-11 01:11:37

A PAVILION FOR THE NON-HUMAN: The underground space of the
Korean Pavilion, where yesterday's land remains intact, is not
easily accessible to humans. However, it has become a place
regularly visited by the animals and insects of Giardini.
If the first floor of the pavilion is a pavilion for humans,
then the underground serves as a pavilion for the non-human.
By respecting the land, the Korean Pavilion has, perhaps
unintentionally, opened itself to the creatures of Giardini.

THE UNDERGROUND THAT HOLDS THE LAND OF THE PAST: The architectural premise and intent to respect the terrain as it was led to the Korean Pavilion's structure, elevated on stilts as if lightly perched above the ground. As a result, unlike most other pavilions, the Korean Pavilion came to have an "underground" space. The underground space beneath the piloti-raised structure preserves the original landscape and soil, maintaining the site's pre-existing topography even after the pavilion's construction.

© Young Yena

Young Yena's '30 Million Years Under the Pavilion' explores the primordial time and space of the Korean Pavilion. Developing through a fictional narrative, the work is the result of collaboration with a biologist, a cultural anthropologist, and an evolutionary sociologist. Dealing with the narrative of the site itself, the installation is situated both underneath the piloti-elevated space of the structure and in the century-old brick building: both the physical origins of the Korean Pavilion. Young Yena summons imaginary guardians of the land, buried millions of years ago, seeking to resonate with a prehistoric temporal axis that far precedes the early history of the Giardini constructed by Napoleon. '30 Million Years Under the Pavilion' awakens a transcendental awareness of the unknown life forms that inhabited the land before the concept of nation-states were invented.

— Chung Dahyoung, Curatorial Essay (2024)

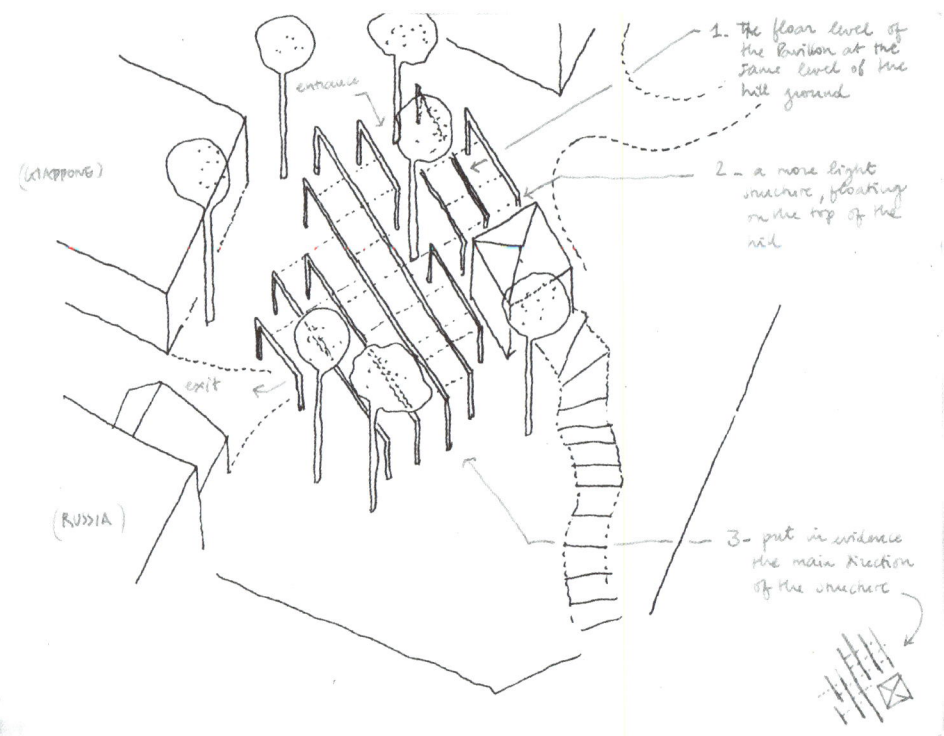

©Mancuso e Serena Architetti Associati. Courtesy of ARKO Arts Archive, Arts Council Korea.

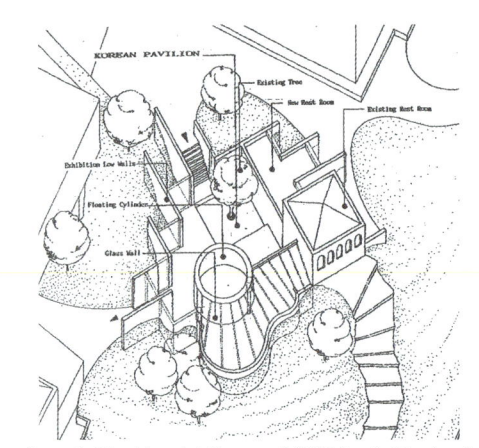

©Mancuso e Serena Architetti Associati. Courtesy of ARKO Arts Archive, Arts Council Korea.

Interior of the Nordic Pavilion, 2023 International Architecture Exhibition
Girjegumpi: The Sámi Architecture Library. Photo: Laurian Ghinițoiu (2023)

Interior of the Canadian Pavilion
Photo: ANDREA PERTOLDEO FOTOGRAFIA, National Gallery of Canada, Ottawa

MAKING ROOM FOR TREES: The most direct design solution for respecting trees as they are is to make room for them within the architecture itself. During the initial design process of the Korean Pavilion, similar approaches were actively considered to fully integrate trees into the interior space. This idea was following the precedent set by relatively recent pavilions of Giardini, representatively the Nordic Pavilion (1962) and the Canadian Pavilion (1958), where trees were integrated into the architectural composition.

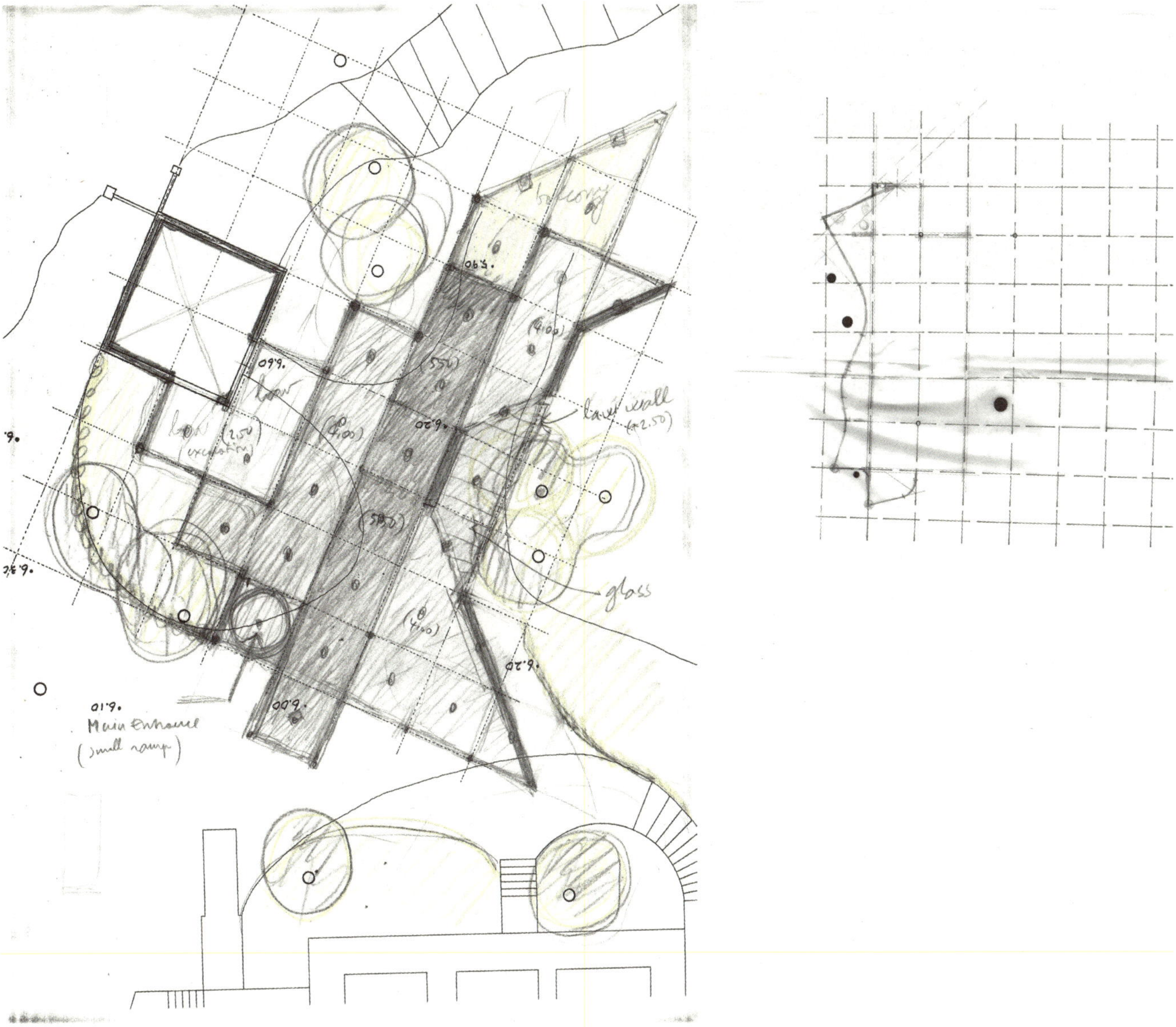

Planar contour studies by Franco Mancuso, regarding the existing trees ©Mancuso e Serena Architetti Associati. Courtesy of ARKO Arts Archive, Arts Council Korea.

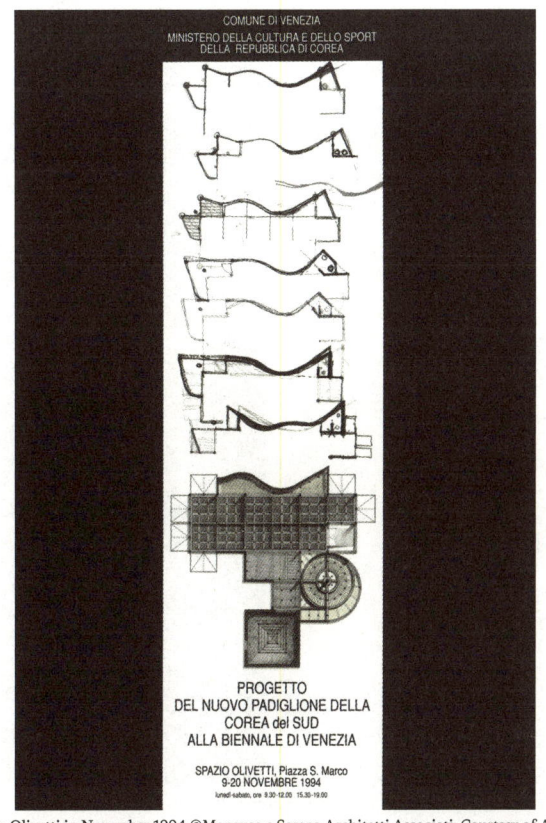

The poster for the exhibition at Spazio Olivetti in November 1994 ©Mancuso e Serena Architetti Associati. Courtesy of ARKO Arts Archive, Arts Council Korea.

©Mancuso e Serena Architetti Associati.
Courtesy of ARKO Arts Archive, Arts Council Korea.

©Mancuso e Serena Architetti Associati.
Courtesy of ARKO Arts Archive, Arts Council Korea.

TREES CARVING OUT THE INFINITE GRID: As the design gradually evolved, the Korean Pavilion moved away from the idea of embracing trees within the interior. Instead, it maintained the principle of an infinitely expandable transparent grid while cutting out its perimeter to accommodate the trees. In other words, defining the contour of the transparent grid with the presence of trees became a key design methodology. This approach appears to be a solution that allowed coexistence with the trees without placing them inside the relatively limited floor space. Initially, the cut-out contours of the grid were defined by straight geometric diagonals, but over time, they began to reflect the free-flowing curves of the trees. The wave-like curves of the walls, shaped by trees' presence, seem to have been ultimately recognized by the architects as one of the defining design elements. This is evident in the poster for the exhibition at Spazio Olivetti in November 1994, which prominently featured the alternative sketches for the undulating wall.

©Mancuso e Serena Architetti Associati.
Courtesy of ARKO Arts Archive, Arts Council Korea.

©Mancuso e Serena Architetti Associati. Courtesy of ARKO Arts Archive, Arts Council Korea.

If the Korean Pavilion, as many critics have described, feels more like a "house" compared to other pavilions, then its approach to trees — keeping them outside the main dwelling area rather than incorporating them within, yet embracing them through large open frames and semi-exterior spaces — feels even more natural, since it aligns with the way traditional Korean houses have historically engaged with trees.

durante la stagione invernale, grazie ad un efficace sistema di riscaldamento.

L'uso del legno - nelle balconate, nella parete ondulata, nelle chiusure esterne - garantisce un'organica relazione con il contesto, anche se l'edificio assumerà due configurazioni distinte nel corso dell'anno; aperto apparirà trasparente e ancor più leggero per la presenza degli slanci costituiti dalle chiusure perimetrali poste in posizione orizzontale, sostenute da cavi; chiuso assumerà l'aspetto di un manufatto che è parte del giardino, perfettamente integrato con l'ambiente arboreo e vegetale circostante.

Con preghiera di pubblicazione/diffusione.
Grazie per la cortese attenzione.

> The use of wood - in the balconies, in the wavy wall, in the external closures - guarantees an organic relationship with the context, even if the building will assume two distinct configurations during the year; open it will appear transparent and even lighter due to the presence of the thrusts constituted by the perimeter closures placed in a horizontal position, supported by cables; closed it will take on the appearance of an artifact that is part of the garden, perfectly integrated with the surrounding arboreal and vegetal environment.

Municipality of Venice's press release regarding the exhibition introducing the Korean Pavilion at Spazio Olivetti (November 1994)
ASAC, Fondo storico, Progetti speciali b. 5874
Courtesy © Archivio Storico della Biennale di Venezia, ASAC

Trees have a life as trees and a life as timber. If we consider their existence as trees to be their first life, then we can say that they live their second life after becoming timber.

— Aya Kōda, "Standing Tree, Lying Tree," *KI* (1992)

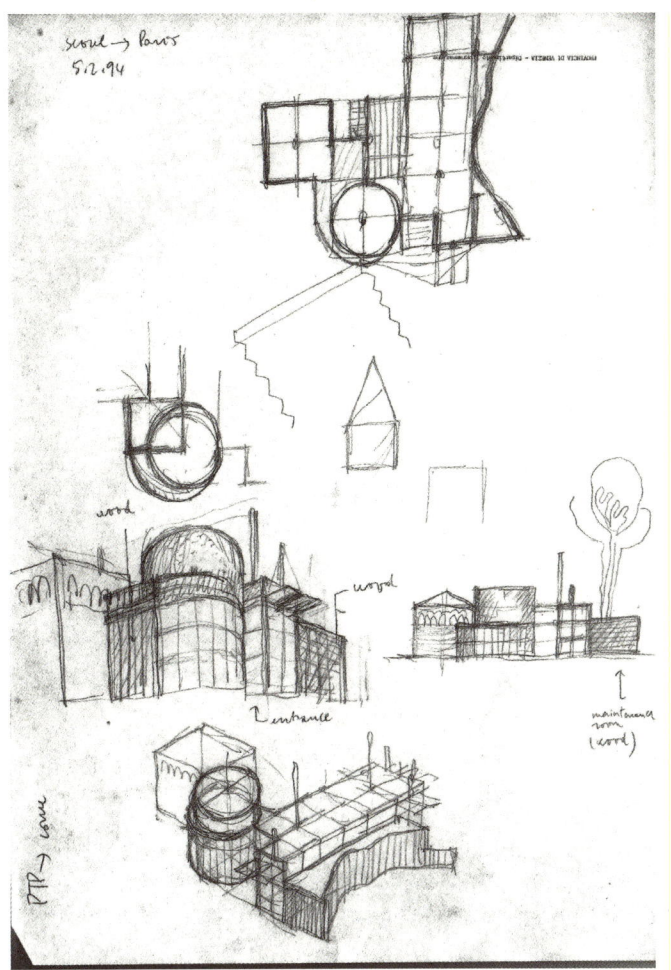

©Mancuso e Serena Architetti Associati. Courtesy of ARKO Arts Archive, Arts Council Korea.

Super Superior Civilizations,
The Austrian Pavilion at the 2024 International Architecture Exhibition

The British Pavilion at the 2024 International Architecture Exhibition

THE HANDS THAT SHAPED THE KOREAN PAVILION, AND THEIR FINGERPRINTS: The organic and undulating contours carved out by the trees became all the opaque elements of the pavilion such as walls and balcony railings. Furthermore, though later replaced with metal, the original kinetic window grilles were entirely made of wood. Hence when fully closed, the opaque areas of the Korean Pavilion were entirely enveloped in wood. In other words, while the pavilion transparently welcomes trees into its space, in its opaque sections, it transforms into wood itself.

Trees as Mediators of National Relations
From a cross-national perspective, trees are more than landscaping elements; they mediate connections between national pavilions and reflect each country's environmental stance and international relations.

Swiss Pavilion's Integration
The Swiss Pavilion also emphasizes the harmony between trees and architecture. Designed by Bruno Giacometti in 1952, the building was constructed while preserving two trees in its front yard and near its walls. As time passed, when these trees fell due to disease, curators preserved their stumps and incorporated them as key elements of exhibitions. This signifies that nature and architecture are not isolated but instead evolve together over time, maintaining a connection to public space.

Trees as Identity and Public Engagement Tools
These examples illustrate that trees are not merely natural objects but act as mediators of identity, solidarity, and public engagement. The differences in how each pavilion incorporates trees are not just about landscaping design but also about shaping relationships between nature, humanity, nations, and public spaces. Over time, trees harmonize with architecture, integrate into spaces in new ways, or convey even stronger messages through their absence.

— Kim Heejung and Jung Sungkyu, Research Essay (2024)

Heechan Park's 'Time for Trees' consists of architectural apparatuses responding to the trees and vegetation of the Giardini—elements deeply embedded in the architectural identity of the Korean Pavilion. Emphasizing the architectural transparency of the pavilion that embraces the surrounding woods, 'Time for Trees' is centrally positioned in the pavilion. The installation, comprising variable installations and drawings, allow visitors to actively capture the natural landscape of the Giardini gardens, including tree shadows. The installation consists of three devices: 'A Shadow Caster,' 'Giardini Travelers,' and 'Elevated Gaze 1995.' Each device serves as a multi-sensorial mediator between the interior and exterior boundaries of the pavilion. Through these interventions, 'Time for Trees' gazes into the trees of the Giardini, the unchanging presence within the ever-shifting context of La Biennale where new people and things come and go every year. 'Time for Trees' ultimately draws attention to the common order, a shared physical context not only for the Korean Pavilion but also for every national pavilion in the Giardini.

— Chung Dahyoung, Curatorial Essay (2024)

© Heechan Park

Though trees have been preserved as part of pavilion developments, this has not necessarily aligned with broader conservation efforts for the park itself. Beyond individual tree protection, which is environmentally important, the overall preservation of Giardini involves maintaining its historical, public, and functional significance.

Beteiligung / Participation, The Austrian Pavilion at the 2023 International Architecture Exhibition / The part between the walls to be bridged
© Clelia Cadamuro

Ultimately, the presence of trees within Giardini is not merely an environmental aspect but a crucial experimental element that explores themes of international cooperation and public engagement. This aligns with the Biennale's transition from a national exhibition model to a more transnational and sustainable framework.

Biennale's Shift to a Transnational and Sustainable Model
As these transformations unfold, Giardini's public nature holds the potential for further expansion. The way nature and architecture are integrated, and how trees are preserved and utilized, transcends environmental concerns to become a key experimental process for fostering global public cooperation. The Biennale must embrace these themes of publicness and sustainability as part of its artistic and social discourse, establishing itself as a platform for open dialogue.

— Kim Heejung and Jung Sungkyu, Research Essay (2024)

Conceptual sketch by Kim Seok Chul ©Mancuso e Serena Architetti Associati. Courtesy of ARKO Arts Archive, Arts Council Korea.

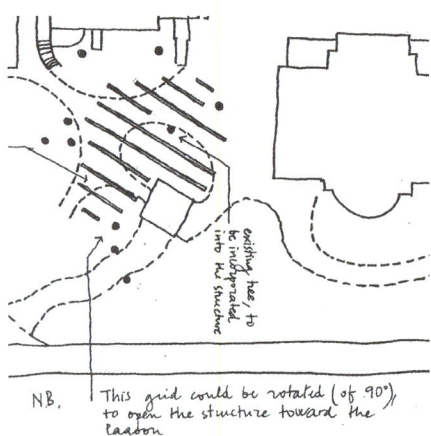

This grid could be rotated to open the structure toward the lagoon.

Early design proposal for the Korean Pavilion by Franco Mancuso ©Mancuso e Serena Architetti Associati. Courtesy of ARKO Arts Archive, Arts Council Korea.

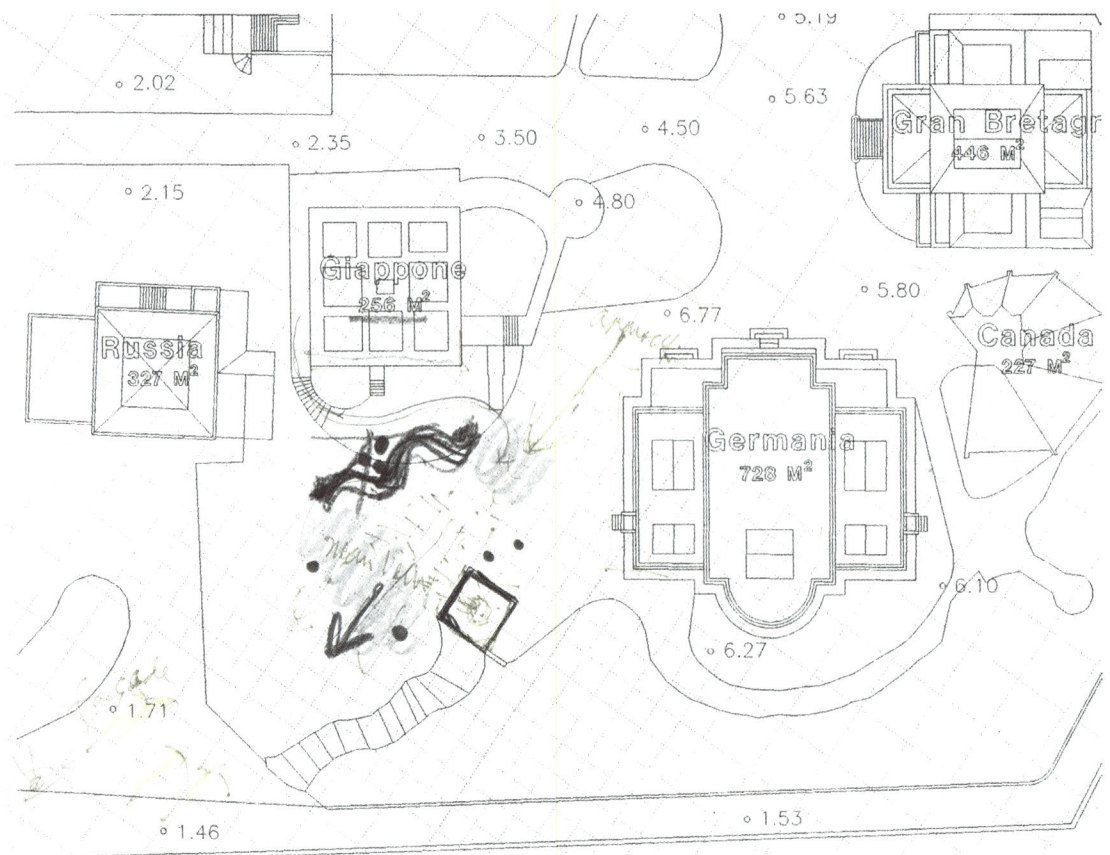

©Mancuso e Serena Architetti Associati. Courtesy of ARKO Arts Archive, Arts Council Korea.

The Korean Pavilion was positioned at a 45-degree angle to the spatial axis of the Giardini, a deliberate choice to provide a better view of the sea.
©Kim Seok Chul. Courtesy of ARKO Arts Archive, Arts Council Korea (Contributor: Kim Seok Woo)

The rooftop of the Korean Pavilion, looking ahead into the Adriatic Sea
©Mancuso e Serena Architetti Associati. Courtesy of ARKO Arts Archive, Arts Council Korea.

©Mancuso e Serena Architetti Associati.
Courtesy of ARKO Arts Archive, Arts Council Korea.

A SHIP BOUND FOR THE BORDERLESS SEA: Though initiated by South Korea, the Korean Pavilion was ultimately envisioned as a structure that would transcend the 38th parallel, bringing reconciliation to the world. Designed with this aspiration, the pavilion does not anchor itself firmly to the ground but instead lifts its heels, poised toward the liquefying ocean. Like a ship's deck, the pavilion's rooftop features two towering masts and numerous halyards, adding to its symbolic meaning. As Kim Hong-hee, who curated a site-specific exhibition at the pavilion in 2003 observed, the raised edge facing the Adriatic Sea resembles a ship docked at a pier, while the undulating walls evoke the waves crashing into the San Marco Basin. The hill where the Korean Pavilion stands was once called Motta di Sant'Antonio, or Sant'Antonio's Mound. In earlier times, it was the "face" of the city that ships encountered when entering Venice. It would be precisely because of this role — as a façade of the maritime city — that Napoleon, after conquering Venice, chose to transform the area into a modern garden in 1807. Now, at the highest point of this historic mound, standing before the sea, the Korean Pavilion turns its bow once again toward the waters that have long dissolved the boundaries between East and West. Like a vessel steering toward the horizon, its architectural axis deliberately defies the parallel alignment of the surrounding national pavilions in Giardini, cutting diagonally across the landscape — charting a course of its own.

Seok Chul Kim
ARCHIBAN, 1-132 Dong Soong-Dong Jong Ro-Gu Seoul,
Korea
Phone 027643072
The 1st of August 1994

Fax To ; Prof. Franco Mancuso
Istituto Universitario Di Architecttura Di Venezia
(041) 523-8121

Thank you for your efforts on the behalf of the Korean Pavilion.

I received your fax dated on the 29th of July 1994 and discussed with the structural engineer.

For the lightness of the building we tried to change the slab of deck plate with concrete to steel panel.(ref. drawing of Detail A) This idea comes from the structure of ship building. I hope to know your idea on this.

The amount of the supports can be reduced to an extend as the drawing shows.

The engineer checked up the drawings and put additional supports which would be located at the meeting point of with the existing building of Edificio in Mattoni. If the existing building could be endurable against the weight that might be loaded when connected with new building, these two additional supports would not be necessary. It would be decided after the structural test of the existing building.

As you recommended, it also seems to me that the visit of Mr. Lee Ung-ho unnecessary at the moment of final session of the Commision. But after the final permission it would be necessary of Mr. lee Ung-ho's visit as a represenative practioner of the Korean Government for the compromise of the official parts. I would like to know your opinion.

Now I am taking into cosideration 5 points you mentioned at the fax dated on 26th July and will visit Venice with the preparation for it between the 29th August and the 3rd of September.

Best wishes and regards
Sincerely,

Seok Chul Kim

A fax letter sent by Kim Seok Chul to Franco Mancuso, August 1st, 1994.
©Mancuso e Serena Architetti Associati. Courtesy of ARKO Arts Archive, Arts Council Korea.

The 'ship' was not only a source of aesthetic inspiration but also a motif for structural design. In his reflections on the construction of the Korean Pavilion, Kim Seok Chul described the process of assembling its steel frame and deck plates as akin to "constructing a ship-like building." The Korean Pavilion as a ship was always destined to set sail — one day departing into the open sea, dissolving borders as it vanishes into the boundless horizon.

Kim Seok Chul and Franco Mancuso sailing on San Marco Basin. ©Mancuso e Serena Architetti Associati. Courtesy of ARKO Arts Archive, Arts Council Korea.

The circular staircase that once guided visitors to the rooftop has since been removed. Now, the rooftop is accessible only via an external staircase. ©Mancuso e Serena Architetti Associati. Courtesy of ARKO Arts Archive, Arts Council Korea.

FORGOTTEN SKY, FORGOTTEN OCEAN: The flat rooftop space of the Korean Pavilion, overlooking the horizon of the Adriatic Sea, was originally designed as a distinct exhibition space, "Gallery 4." Had it not been intended for public access, such distinctive and delicate architectural elements would not have been necessary. At the time of completion, a small circular staircase was placed at the center of the Cylinder Hall, allowing visitors to ascend to the deck and encounter the sea and sky in a dramatic way. However, as the staircase was seen as an obstruction to the exhibition space, it was soon removed. Over time, the ocean and sky, once actively embraced by the pavilion, gradually faded from memory.

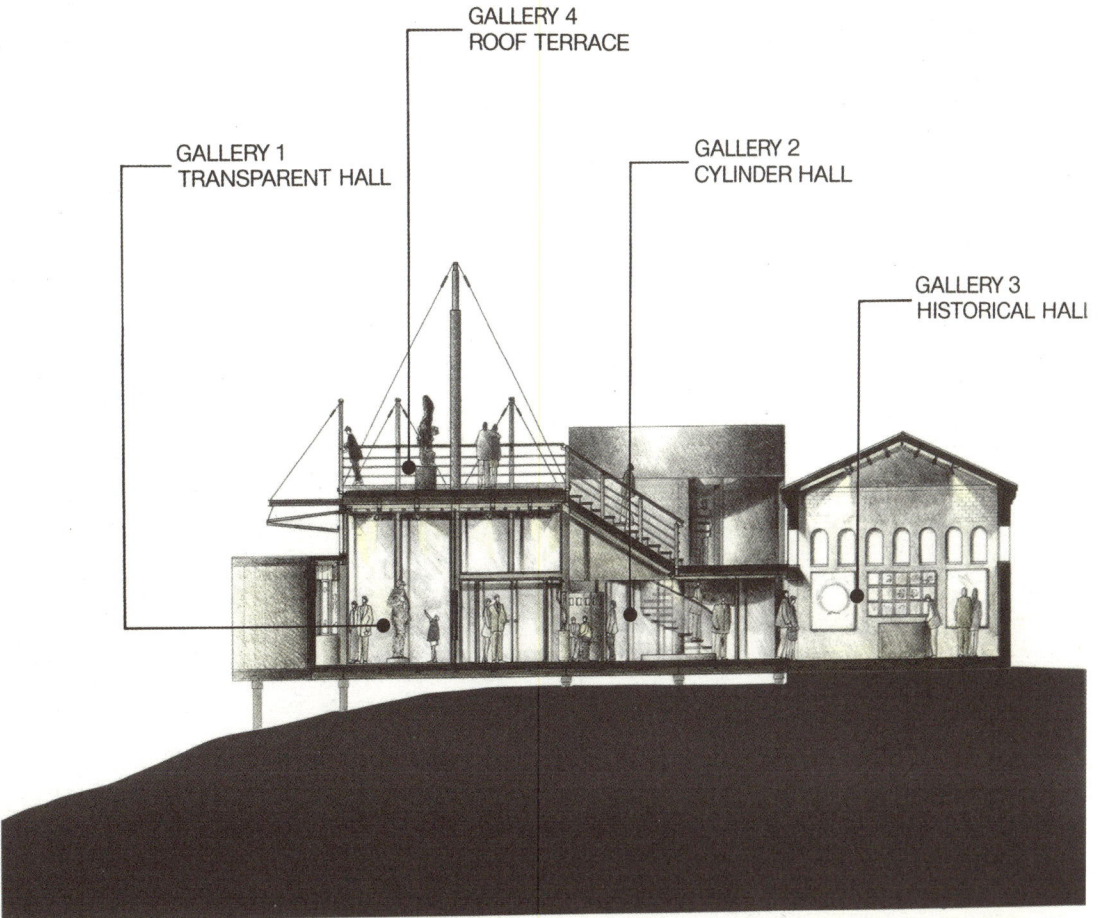

©Kim Seok Chul. Courtesy of ARKO Arts Archive, Arts Council Korea (Contributor: Kim Seok Woo)

Kim Hyunjong's 'New Voyage' is installed on the rooftop, a unique but overlooked space of the Korean Pavilion. Access to rooftop spaces are usually impossible or limited in national pavilions of the Giardini. 'New Voyage' revitalizes the rooftop of the Korean Pavilion, originally conceived as an exhibition space, while urging unknown future possibilities. Instead of looking inward toward the Giardini, 'New Voyage' is oriented outward, facing the Adriatic Sea. This positioning draws inspiration from the pavilion's architectural form, which resembles a ship with sails set toward the sea. By reopening the space to visitors in tandem with the installation, the rooftop transforms itself into a space of hospitality, an observatory for all. 'New Voyage' invites visitors to reconsider the pavilions' relationship with their shared natural elements—the sky and the sea—that transcend national boundaries and connect the pavilions of the Giardini.

— Chung Dahyoung, Curatorial Essay (2024)

©Mancuso e Serena Architetti Associati. Courtesy of ARKO Arts Archive, Arts Council Korea.

"Toad, Toad, wake up. It is May now."
"What?" said Toad.
"Can it be May so soon?"
"Yes," said Frog.
"Look at your calendar."
Toad looked at the calendar.
The May page was on top.
"Why, it is May!" said Toad as he climbed out of bed.
Then he and Frog ran outside to see how the world was looking in the spring.

— Arnold Lobel, *Frog and Toad are Friends* (1970)

A view of the exhibition introducing the Korean Pavilion project, held at Spazio Olivetti in Piazza San Marco, Venice (November 1994). On the left side, a model of the pavilion is displayed. ©Kang Suk-Won. Courtesy of ARKO Arts Archive, Arts Council Korea (Contributor: Kim Dong Jin).

WORKS

EXHIBITION DRAWINGS

Kim Giseok

ROOFTOP

- Lee Dammy, Overwriting, Overriding
- Young Yena, 30 Million Years Under the Pavilion
- Heechan Park, Time for Trees
- Kim Hyunjong, New Voyage
- Exhibition Devices

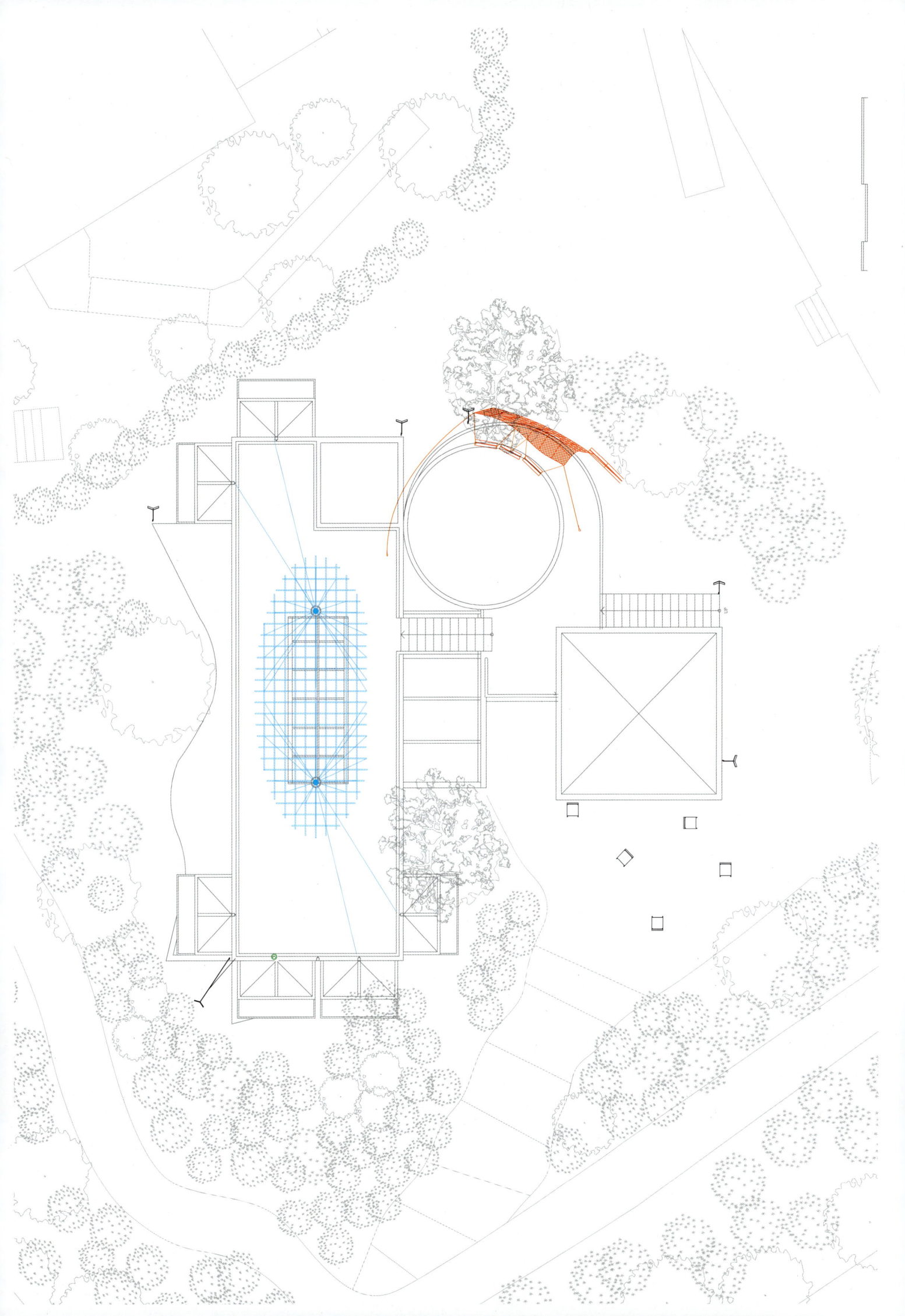

1F
— Lee Dammy, Overwriting, Overriding
— Young Yena, 30 Million Years Under the Pavilion
— Heechan Park, Time for Trees
— Kim Hyunjong, New Voyage
— Exhibition Devices

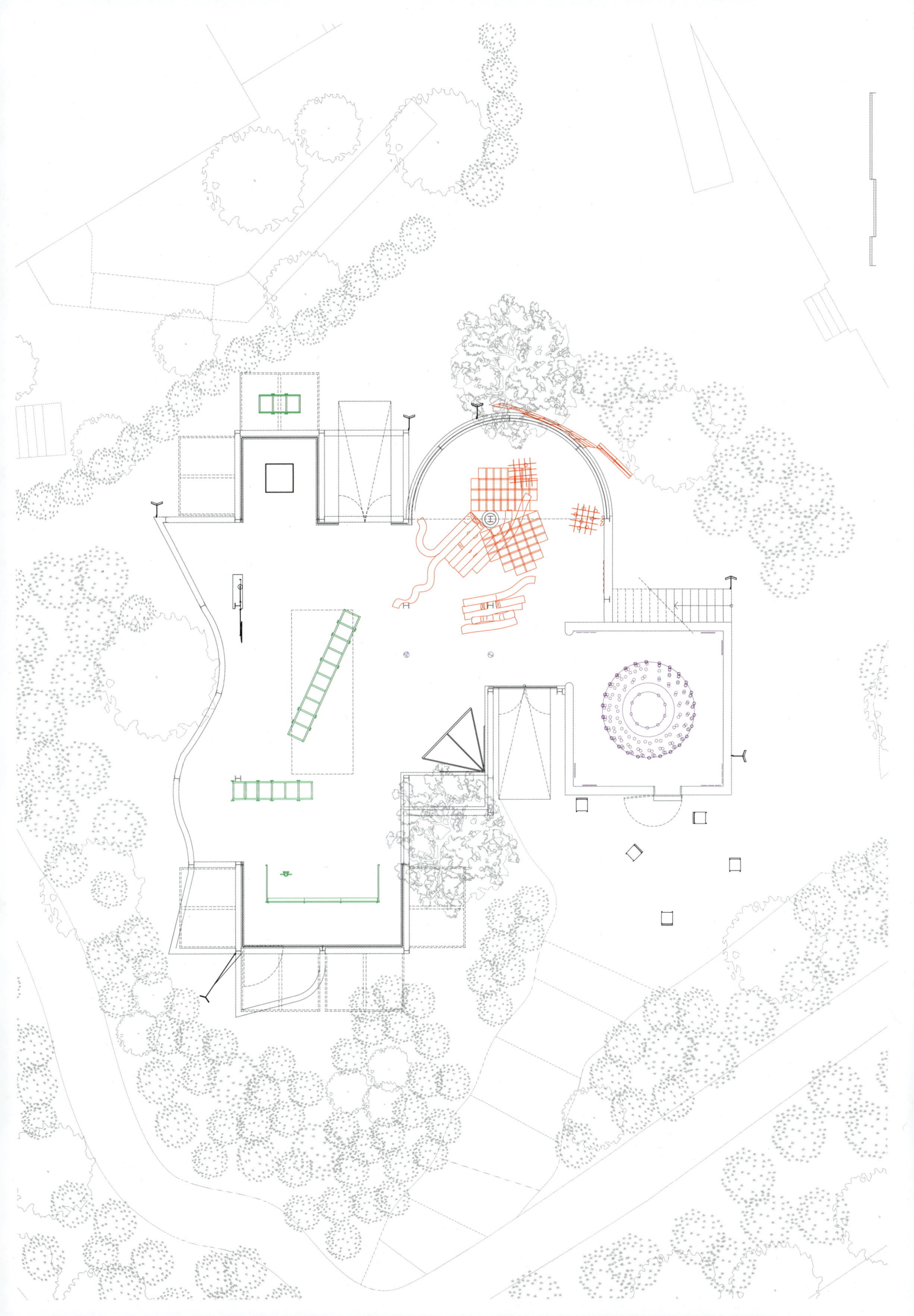

UNDERGROUND

- Lee Dammy, Overwriting, Overriding
- Young Yena, 30 Million Years Under the Pavilion
- Heechan Park, Time for Trees
- Kim Hyunjong, New Voyage
- Exhibition Devices

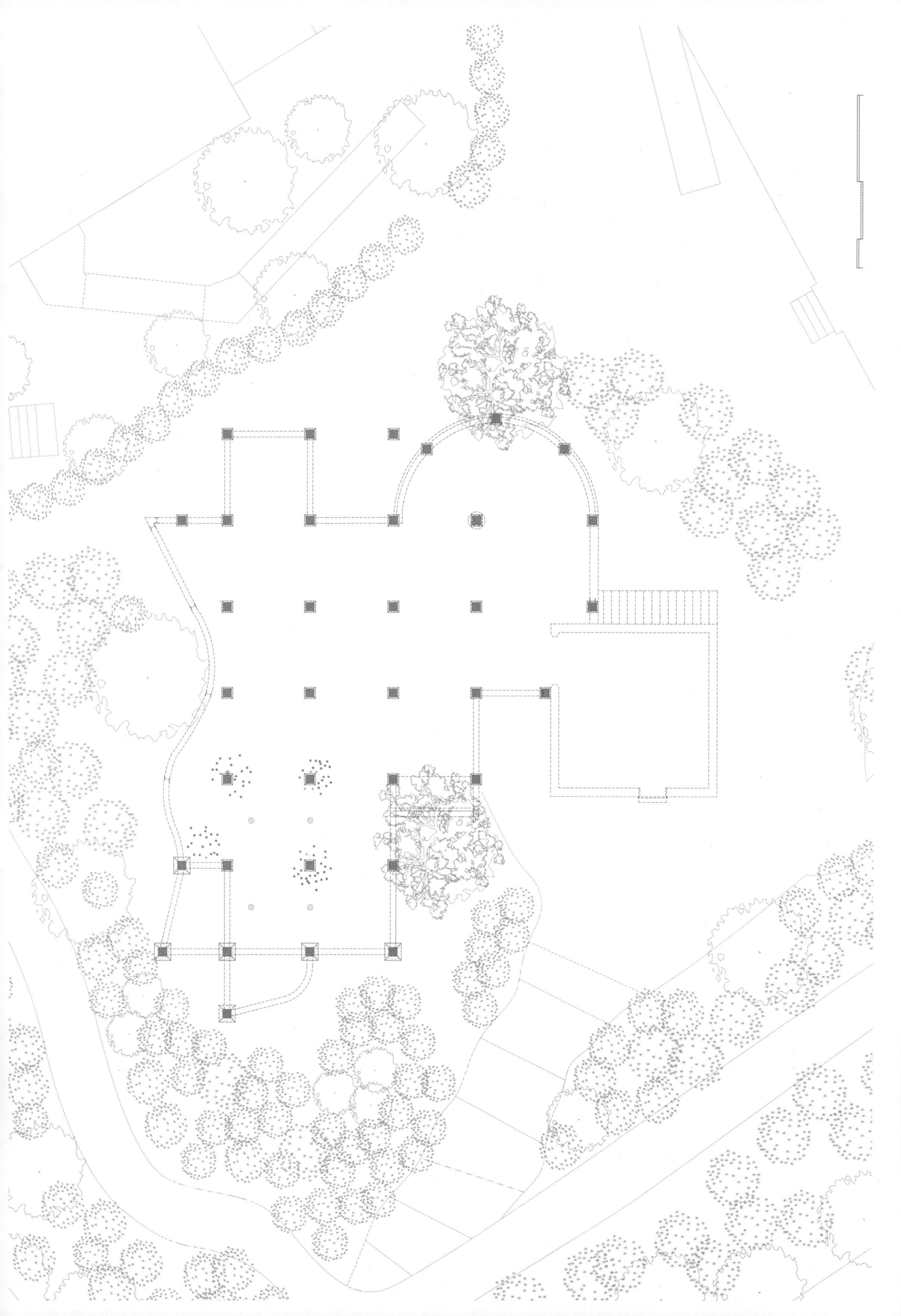

SECTION

- Lee Dammy, Overwriting, Overriding
- Young Yena, 30 Million Years Under the Pavilion
- Heechan Park, Time for Trees
- Kim Hyunjong, New Voyage
- Exhibition Devices

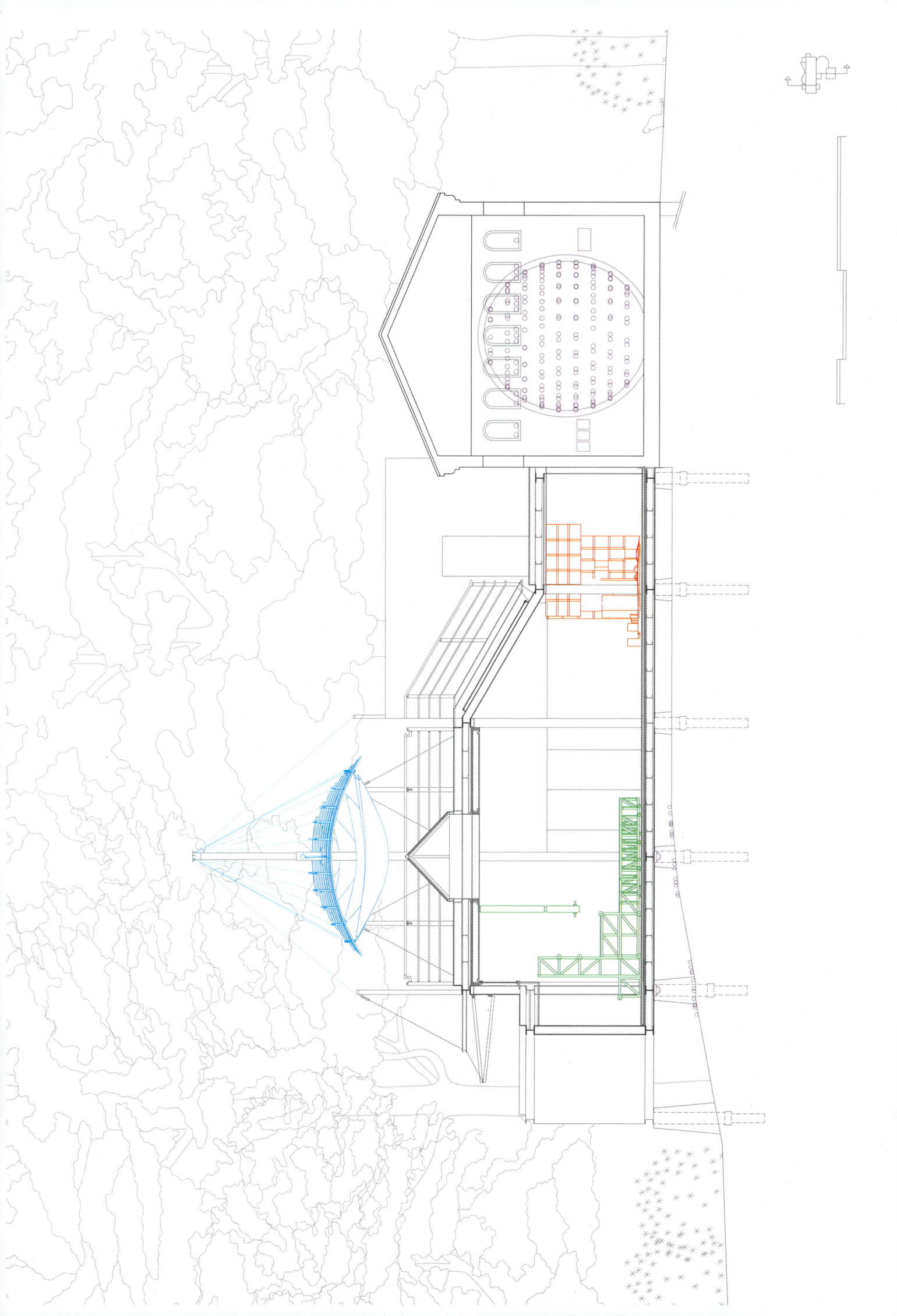

Lee Dammy

Marginal

The Last Pavilion in Giardini: Overwriting the Layers of Identity

January 20, 2025 By Giuha Bianchi

In the paradoxical, permanently temporary setting of the Giardini della Biennale, where national pavilions narrate stories of identity and pride, the Korean Pavilion is poised to turn the page. With an ambitious vision to transcend its historical role as a national pavilion dedicated to showcasing South Korea, it now aspires to become a permanently open venue—an inclusive space where boundaries dissolve and dialogue flourishes.

Venice Biennale Korean Pavilion

Erected in 1995 as the 26th OVERWRITING OVERRIDING in the Giardini, the Korean Pavilion has served as a pivotal platform for Korean art on the global stage. Its presence marked a turning point for South Korea, whose participation in the

OVERWRITING, OVERRIDING
Writing more and more of the future, the present, and the past

---

OO PAVILION
In 2025, as the Korean Pavilion at the Venice Biennale marks its 30th anniversary, it makes a decisive move toward a new future. Shedding its limited role as a national pavilion opening for periodic Biennale, it fully embraces its original vision, transforming into an ever-open public space — now simply the ( ) Pavilion. To facilitate this transition, the government announces an international design competition, calling for proposals that subtly strip away overt markers of Korean identity. To ensure fairness, the jury remains anonymous.

Architect Dammy Dami Lee embarks on a journey to reimagine the identity of this space, one that was once the Korean Pavilion but soon will not be. Instead of nationalism, she seeks an alternative sense of belonging. Her inquiry begins with her most immediate point of connection — the fact of her own existence as a woman. But her search soon veers off course: she becomes absorbed in the presence of the honey locust tree, a silent witness to the pavilion's history, and Mucca the cat, who has inherited this land through generations. The more she listens, the more she is drawn away from her original task, her focus scattering like loose threads.

She tries to mediate their voices, translating them into design. But architecture resists direct translation. She realizes that even interpreting her own identity — as a woman, as an architect — is an intricate and uncertain process.

DOES IDENTITY EVEN MATTER?
Yet paradoxically, the more she attempts to erase the concept of a nation, the more Korea emerges — vivid, fragmented. Design is no longer about the pure elements of architecture but becomes what was once considered peripheral — woven mats, faded wallpaper, embroidered textiles, ornamental motifs, exaggerated facades, and the space of dreams and desires. Architecture dissolves into traces of what was abandoned, into fleeting moments, into reveries of the present.

CAN SHE WIN THE COMPETITION WITH THESE ELEMENTS?
Then, just as she lingers before an unanswerable question, Tangun as a Scythian King, a robot figure from Nam June Paik's universe, appears. "I am the Mother of the Korean Pavilion," they declare, placing an iced Americano into her hands.

Condensation gathers on the cold glass, leaving a damp stain on the napkin where she has been sketching. Watching the stain spread, Dammy Dami Lee recalls the moments when she had become the honey locust tree, become Mucca the cat, become herself — immersed in conversation, in imagining, in speculation. And in this quiet shift, she understands perhaps this, too, is what architecture is about.

Biennale began in 1986 without an independent exhibition space. The dramatic shift came in 1993 when Nam June Paik, representing Germany, won the prestigious Golden Lion, sparking a collective call for a pavilion that could embody South Korea's cultural presence.

Designed by Korean architect Kim Seok-chul and Italian architect Franco Mancuso, the pavilion opened as a testament to collaboration and vision. Kim's design, inspired by the ideals of transparency and openness, hinted at a future where the pavilion might transcend its episodic utility. That future has now arrived.

As the pavilion approaches its 30th anniversary, the Korean government has taken a bold step to redefine its purpose. Suspending physical expansion proposals considered since 2017, the focus has shifted to reimagining its role within the Giardini. This transformation is spearheaded by Dammy Dami Lee, a rising architectural talent chosen through a public competition. Under the theme 'OO Pavilion - Post-nationalism Korean Identity', Lee's project, titled 'Overwriting, Overriding', embraces the challenge of layering new meanings onto a structure rooted in history.

Dammy Dami Lee

The architect explains that her work is not an attempt to provide a closed conclusion but rather a speculative stage for performative role-playing, akin to an ongoing theatrical exploration. She adds that discovering and imagining new characteristics beyond nationality compels us to reexamine this place and ourselves, shifting the project away from being confined to a purely white, transparent, and pristine structure. Instead, it delves into devices that respond, reflect, contaminate, and connect—fragile, layered, and perpetually temporary shells, embodying the fluidity and complexity of identities.

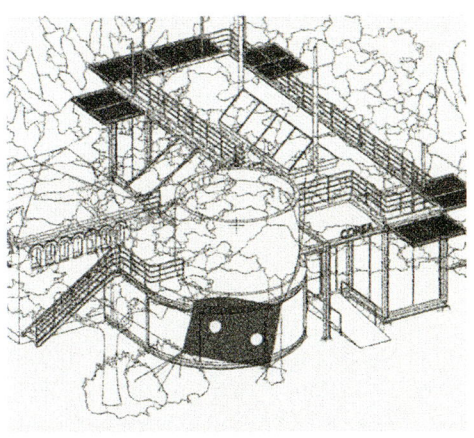

Overwriting, Overriding

Beginning with the intimate observations of herself as an Asian female architect, Lee's work gathers and resonates with diverse yet previously unheard voices. Her words capture the spirit of the project *"To veil and unveil, to invite and redefine, and to both represent and misrepresent—it is both a joy and a responsibility, exploring the possible and impossible."* she explained. She also quoted David Lynch, who recently passed away, saying, *"The Individual is cosmic."*

Over time, her process will unfold, not as a fixed or definitive answer, but as an open invitation for viewers to interpret its meaning personally.

Dear Dammy Dami Lee,

Hello. I'm writing this letter after reading the article published in Marginal. I am "the tree in front of the cylinder." I have many names, but people often call me this, so I thought you might recognize me by it as well. I heard you'll be reimagining the Korean Pavilion, and I'm curious and excited to see what you'll create. When the building was first erected, I was still young. Back then, I found joy in simply watching the building rise beside me and seeing people come and go. Now, thirty springs have passed. The traces of the wind between the cylinder's glass and my trunk, as well as the way my branches stretch out with care to avoid the rooftop's edge, have all become part of my form. The Korean Pavilion has become a part of me, and I, a Gleditsia triacanthos, or honey locust tree, have become a part of the pavilion. Perhaps those who waited for their friends beneath my shade in front of the cylinder might feel the same way.

So this time, I hope you'll make a beautiful proposal—not just for the pavilion I've watched over all these years, but for me as well, now that I, too, have become a part of it.

Carrying the breeze of Giardini,

The Tree

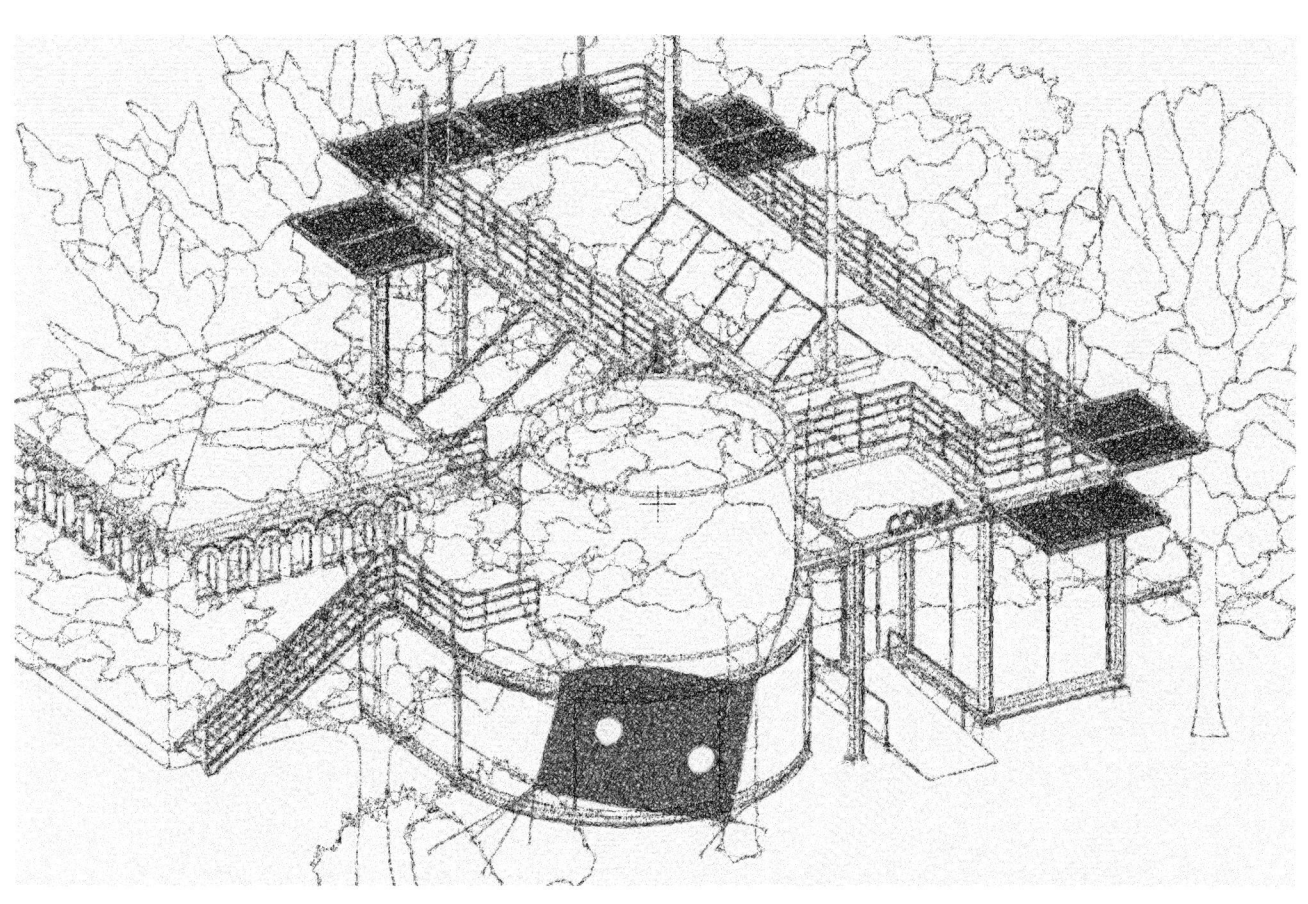

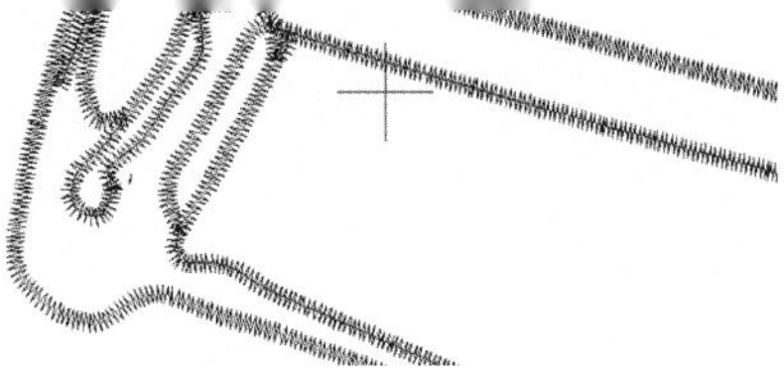

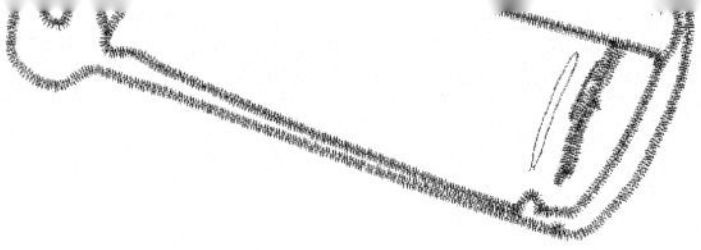

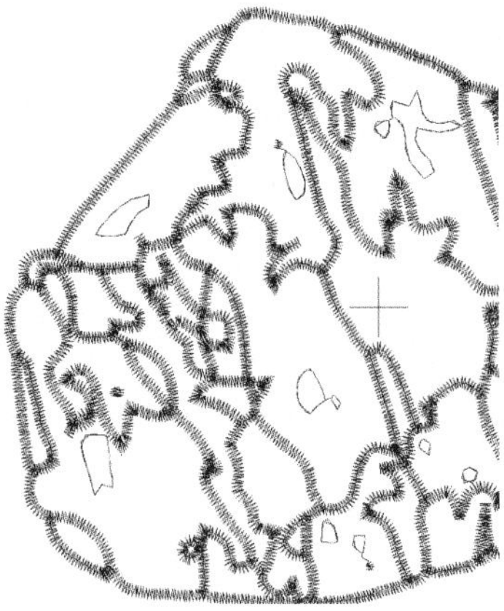

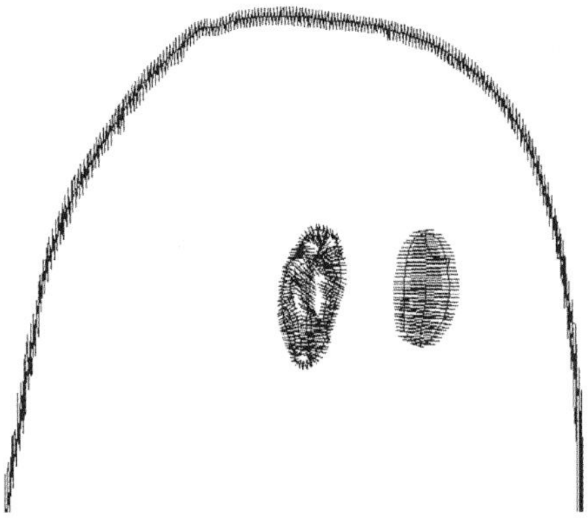

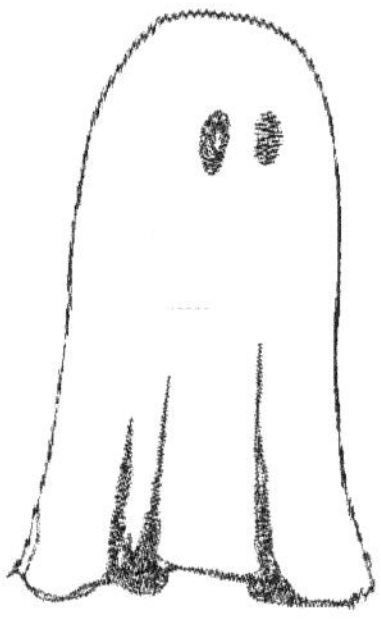

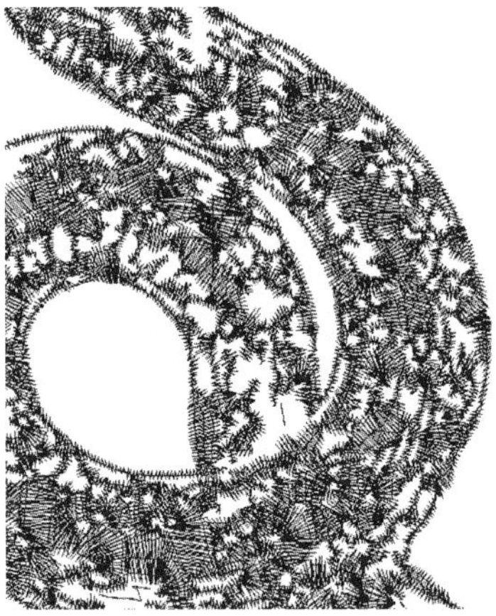

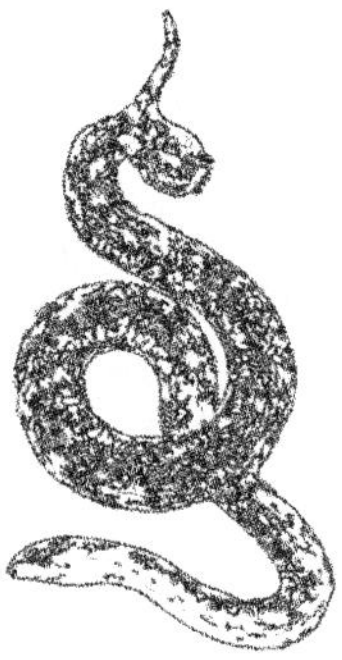

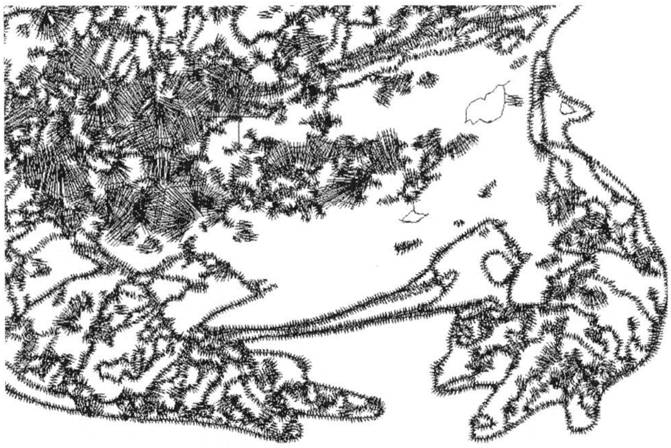

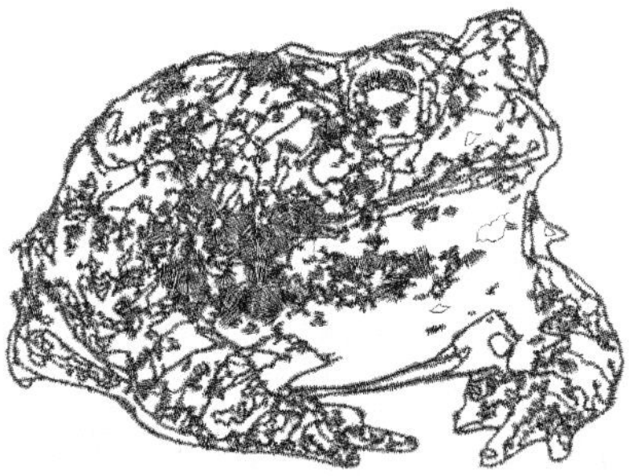

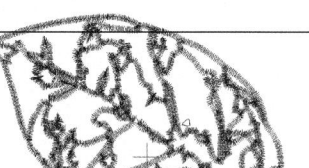

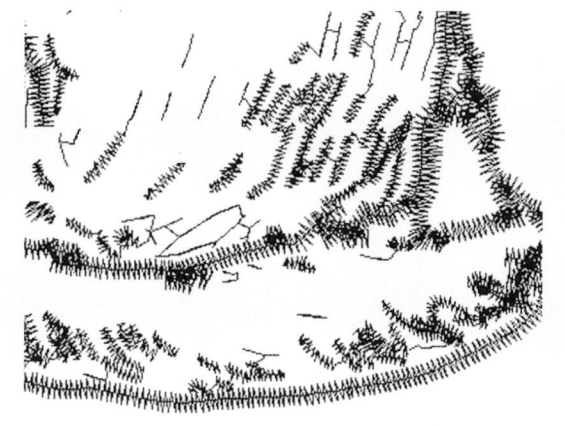

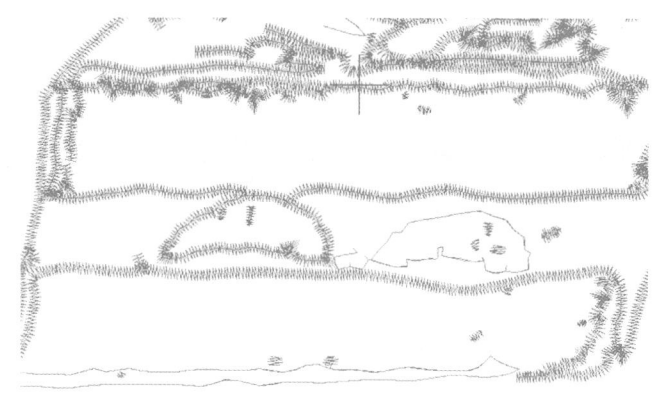

y a quelqu'u
loin poi
떤 사람이 있
멀리서 온 ㅁ

Il y a quelqu'une point
loin point
어떤 사람이 있어요 다침표
멀리서 온 다침표

ㄴ 일으로 읊조린
반대편에 새로운

OVERWRITING, OVERRIDING

당신은 읽고
당신은 인으로 읊조린다.
당신의 반대편에 새로운 상태로,

124 DAMMY

OVERWRITING, OVERRIDING

WORK INFORMATION
- A: Embroidery on sewn nylon mesh fabric, Stainless steel wire and weights, Dimensions variable, 2024
- B: Printed letters and punched holes on layered hanji (Korean indigenous paper), Dimensions variable, 2024
- C: Comforter and pillows filled with cotton, Buttons and embroidery on covers made from various fabrics, Collected items like twigs and fallen leaves, Dimensions variable, 2024
- D: Furniture made from stainless steel rods, Objects such as cups, napkins, stones and wood chips, Dimensions variable, 2024

CREDITS
- Installation Assistance: Kim Yujin
- Metalwork and Installation: Injure
- Hanji Paperwork: Park Eunwook
- Sewing Consultation: Cha Jeongsun
- Document Provision: Paik Jongkwan, Choi Yunha, Choi Eunji, Ho Kyungyoon
- Supported by: Kim Sunam

Young Yena

30 MILLION YEARS UNDER THE PAVILION
Where the permanent meets the temporary in a tapestry of architectural storytelling

---

TEMPORARY ARCHITECTURE AND THE PERMANENT EXHIBITION BENEATH
In 2025, the Korean Pavilion at Giardini della Biennale will celebrate its 30th anniversary. Originally designed with its eventual dismantling in mind, the pavilion has defied expectations by enduring for over three decades. In South Korea, buildings of this age are commonly considered at the end of their typical lifecycle, often leading to redevelopment or reconstruction. As a pavilion - a structure intended for specific, often temporary purposes - the Korean Pavilion embodies the tension between permanence and temporality, challenging conventional notions of lifespan and function.

Contrasting the pavilion's now-permanent temporary structure, a permanent exhibition will be installed beneath its floor. This hidden display will only be fully revealed when the pavilion itself is dismantled in the future. This deliberate concealment invites visitors to ponder the unseen layers of history and architecture that lie beneath the surface of our seen world.

UNEARTHING THE MYSTERY: AN ARCHAEOLOGICAL SITE
The exhibition transforms the Korean Pavilion as an archaeological site in Venice, where, in 1993, Korean artist Young Yena made a startling discovery while preparing her outdoor sculpture installation for the German Pavilion. On the site where the Korean Pavilion now stands, Young uncovered mysterious forms lying on the soil, which initially appeared to be ancient clay figurines. However, further examination revealed that these were not human-made artifacts but the remains of an unprecedented hominid species — *Nanogyna acephala* ("small headless woman" in translation).

Remarkably, much like trovants—stone-like formations known for their ability to grow and "live" under the right conditions— *Nanogyna* defies expectations with its slow yet continuous growth and adaptation. This astonishing discovery has perplexed scientists and anthropologists for decades, challenging traditional notions of human evolution and archaeology, and forcing researchers to reconsider the boundaries between the living and the nonliving.

NANOGYNA ACEPHALA N. SP.

Standing at just 5-18 cm tall, *Nanogyna acephala* has a humanlike body with pronounced breasts and pelvic structures but lacks a head. Its upper torso appears sealed, leading some to speculate that it may rely on torso-based sensory perception or non-neural communication, challenging conventional biological understanding.

Communication seems to occur through frequencies and a rudimentary language of beeswax balls, where the number and composition of the balls likely convey meanings. When touched by water, *Nanogyna* produces sounds similar to crickets, likely due to specialized structures that generate vibrations. These sounds could serve as a form of communication, amplifying frequency-based signals. Water may also help transmit these vibrations over distances, enabling *Nanogyna* to interact with its surroundings.

Genetic analysis reveals that the Hox genes 1-4, typically found in the head region of modern humans, are located in *Nanogyna*'s sexual organs, hinting at an unusual evolutionary path. *Nanogyna* exhibits an extraordinary reproductive ability through parthenogenesis, a form of asexual reproduction where offspring are produced by females without the need for male fertilization. This process seems to be linked to their high reproductive potential, and is driven by hormonal mechanisms similar to Xenopus frogs. Just as hormones in the urine of pregnant women influence ovarian development and trigger egg-laying in Xenopus, *Nanogyna* may rely on hormonal cues allowing it to thrive in specific environments.

THE ENVIRONMENT OF *NANOGYNA*: WHY GIARDINI IN VENICE?

The discovery of *Nanogyna* in Giardini, Venice, is no coincidence. This unique location, with its rich biodiversity and proximity to water, may have provided the perfect conditions for the species. The area's microclimate, combined with historical human activity and natural ecosystems, likely supported the species' unusual physiology and reproductive needs. Some theorize that *Nanogyna*'s adaptation to this environment allowed it to survive in isolation, preserving its fossilized remains until their eventual discovery.

AN EVOLUTIONARY PUZZLE

The discovery of *Nanogyna* has sparked debate about its place in the evolutionary tree. Unlike traditional hominids, *Nanogyna* represents an entirely alternative evolutionary path. Some speculate that *Nanogyna* was an evolutionary "experiment," relying on non-neural cognition and environmental adaptability instead of sensory organs. The strange, spiraled growths in place of arms suggest it may have been semi-aquatic, using plant-like appendages for nutrient absorption or movement.

These raise questions that extend beyond science. What if *Nanogyna* represents a parallel lineage to humans, one that evolved under completely different circumstances? What does its existence reveal about the fleeting nature of evolution and life's experiments?

REWRITING THE NARRATIVE OF PERMANENCE

The story of *Nanogyna* forces us to reconsider what we know about evolution, permanence, and the narratives we construct about the past. Just as the pavilion itself — designed for impermanence — stands as a reminder of the tension between the ephemeral and the enduring.

Through the lens of *Nanogyna*, the exhibition transforms the Korean Pavilion into a stage for speculative archaeology, inviting visitors to uncover the hidden stories that lie beneath the surface of our world — and to question how much of history, like the *Nanogyna*, remains unseen.

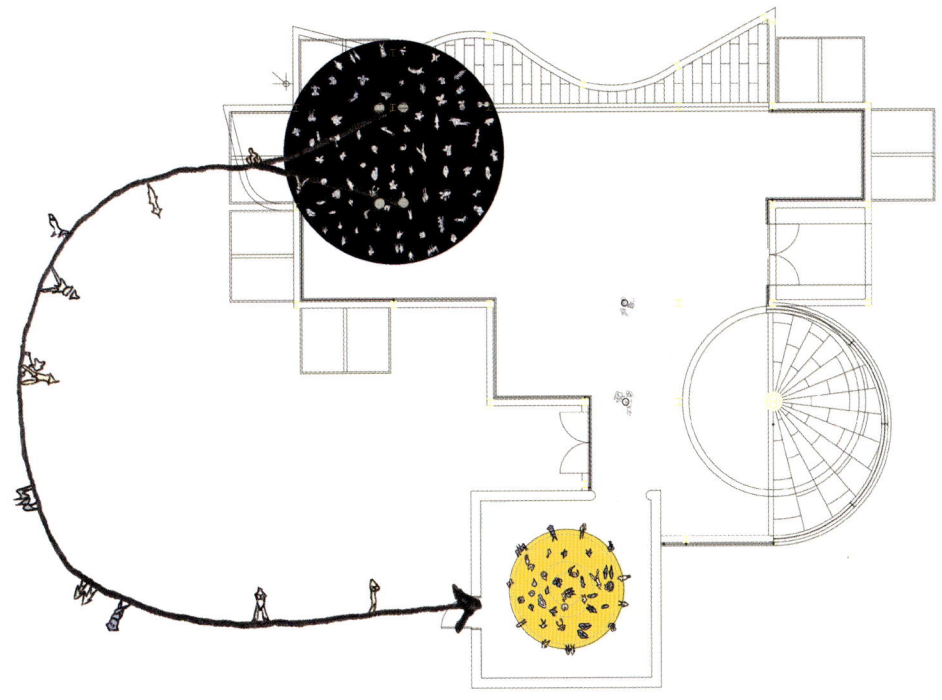

↑ Exhibition plan   ↓ Exhibition section

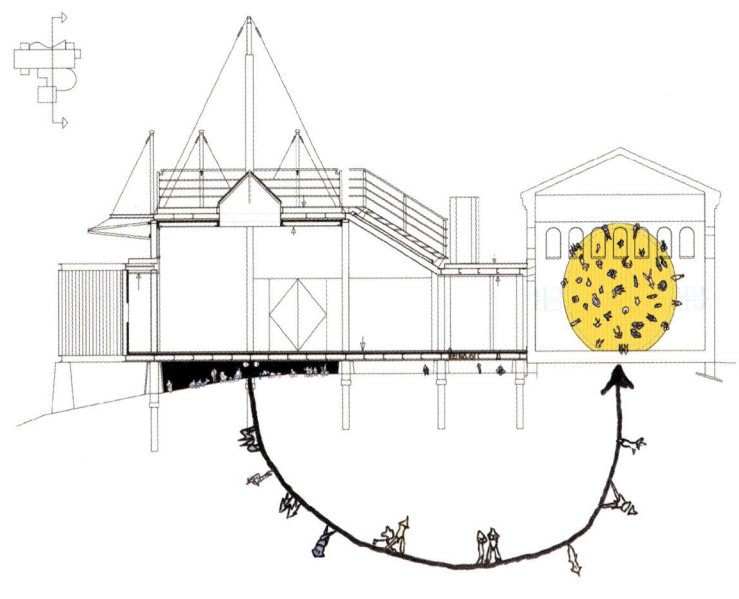

↓ Found *Nanogyna*

134   YOUNG YENA

↓ Found *Nanogyna*

30 MILLION YEARS UNDER THE PAVILION

↑ The main found site of *Nanogyna*.
Giardini della Biennale, Venice, 1993

↓ Young Yena in front of her work for
the German Pavilion, 1993

↑ In front of German pavilion.
Found *Nanogyna*, 1993

↓ *Nanogyna*'s potential language and numbers made of beeswax

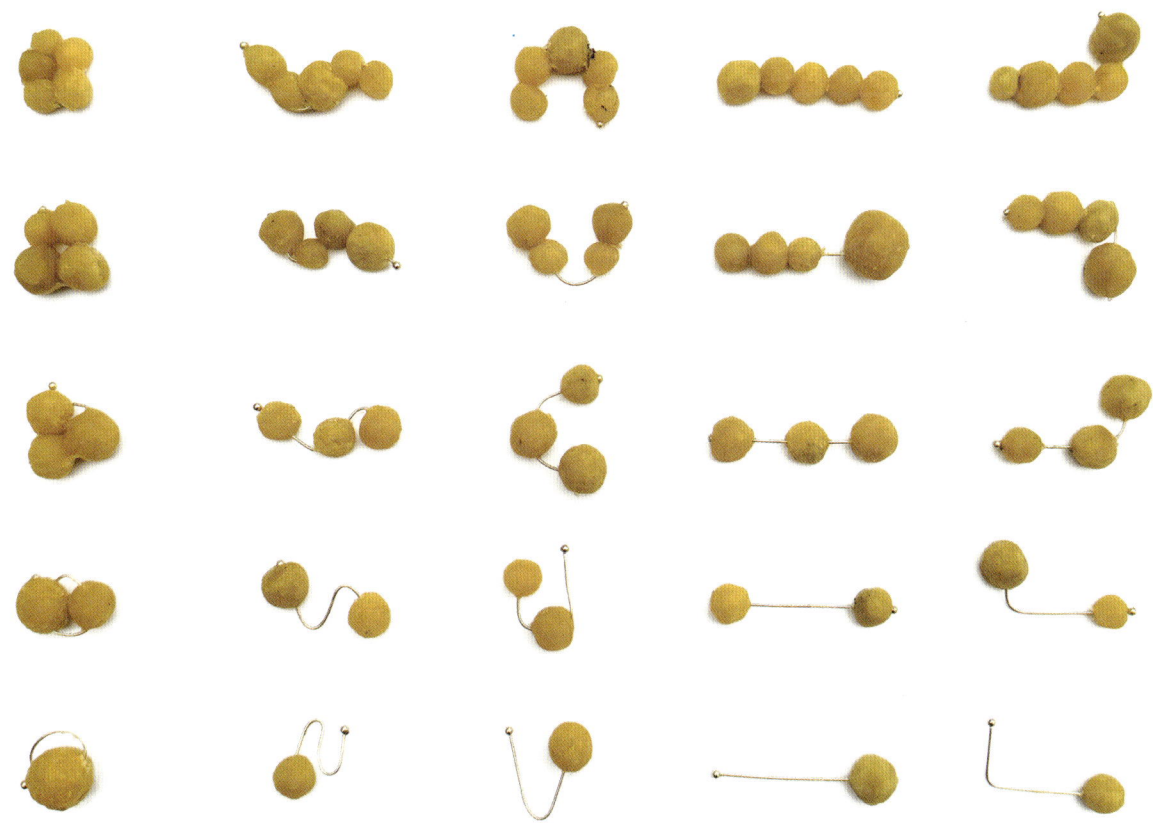

↓ Inventory of found *Nanogyna* (page 3), Young Yena, 1993

PAGE. 3		SUBJECT: Nanogyna	where are the Heads?		Museum für Naturkunde der Universität Berlin.	
1.	2.	3.	4.	5.	6.	7.
NUMBER	DRAWING	SPECIFICATION	GENOTYPE	SUBTYPE	COLOUR	SIZE
N.PE-BR-031		Hole in the middle Front & Back Not going through	Nanogyna acephala	perforata	Orange	W: 40 L: 90 H: 40
N.DI-BR-032		Three separate parts. with going through Holes.	Nanogyna acephala	dissecta	Orange in ash	W: 140 L: 50 H: 35
N.DI-BR-033		Four separate parts. with going through Holes.	Nanogyna acephala	dissecta	Orange in ash	W: 150 L: 40 H: 20
N.ME-PA-034		Mixture of colours of Beeswax balls in a thin bowl.	Nanogyna acephala	melliglobosa	Yellow grey black	W: 60 L: 60 H: 20
N.SP-PA-035		N.SP-PA-035 and 036 seems to be related.	Nanogyna acephala	sphaerocorpus	Sandy gray	W: 50 L: 60 H: 70
N.SP-PA-036		The sphere balls have developed from the bottom to the top	Nanogyna acephala	sphaerocorpus	Sandy gray	W: 60 L: 50 H: 90
N.BA-PA-037		Holes going through It can be also perforata.	Nanogyna acephala	batrachoides	Beige in ash	W: 75 L: 105 H: 80
N.EC-PA-038		Dark Brown horn. Black Sandy grain with holes	Nanogyna acephala	echinosoma	Brown in ash	W: 60 L: 60 H: 90
N.BA-PA-039		It seems there were thin pins	Nanogyna acephala	batrachoides	Purple in ash	W: 90 L: 135 H: 25
N.BI-PA-040		It can be geminata. TOP SIDE	Nanogyna acephala	bifundalis	Purple in ash	W: 60 L: 80 H: 25
			YEAR: 1993	NAME: YOUNG		

↓ Inventory of found *Nanogyna* (page 21), Young Yena, 1993

PAGE. 21 SUBJECT: Museum für Naturkunde
                                                      der Universität Berlin.

1. NUMBER	2. DRAWING	3. SPECIFICATION	4. GENOTYPE	5. SUBTYPE	6. COLOUR	7. SIZE
N.HE-PA-211		very long arms wrapping legs / Side view	Nanogyna acephala	helicoides	Brown in ash	W: 40 L: 65 H: 45
N.SP-PA-212		Side view / potentially fell off?	Nanogyna acephala	sphaerocorpus	Brown in ash	W: 40 L: 60 H: 15
N.PE-PA-213		slit approx. 7mm	Nanogyna acephala	perforata	Brown in ash	W: 40 L: 65 H: 40
N.PI-PA-214		Frequency Sound Source? / Front Back	Nanogyna acephala	pisciformis	Purple in ash	W: 40 L: 90 H: 20
N.SP-PA-215		four breasts between legs	Nanogyna acephala	sphaerocorpus	Brown in ash	W: 50 L: 40 H: 60
N.PE-PA-216		slit only on front approx. 8mm	Nanogyna acephala	perforata	Brown in ash	W: 35 L: 95 H: 30
N.SP-PA-217		thick rounded than n.197 / Side view Top view	Nanogyna acephala	sphaerocorpa	Brown in ash	W: 35 L: 40 H: 35
N.BA-PA-218			Nanogyna acephala	batrachoides	Purple in ash	W: 45 L: 100 H: 25
N.AB-PA-219		arms missing or fell off or not growing? / Top view	Nanogyna acephala	abrachium	Purple in ash	W: 35 L: 130 H: 25
N.PE-PA-220		slit only on front approx. 10mm	Nanogyna acephala	perforata	Purple in ash	W: 35 L: 120 H: 25

YEAR: NAME: YOUNG

↓ A fax letter regarding the discovery of *Nanogyna* in the future site of the Korean Pavilion, 1993

8. JAN '93 12:18 VON WLMK UK 4400 MS AN 9415210038 SEITE 001

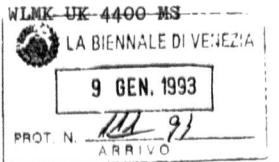

LANDSCHAFTS-
VERBAND
WESTFALEN-LIPPE

Westfälisches Landesmuseum für Kunst und Kulturgeschichte
Domplatz 10 4400 Münster

Al Sig.
Achille Bonito Oliva
– Biennale di Venezia –
Fax. 0039/41/521 00 38

**Westfälisches
Landesmuseum für
Kunst und
Kulturgeschichte**
Domplatz 10
Vermittlung: (02 51) 59 07 01
Durchwahl: (02 51) 59 07 -
Telefax-Nr: (02 51) 59 07 210
TX 892835 lewel d

XX-1128
Aktenzeichen im Antwortschreiben bitte angeben

MUENSTER 7. I. 1993

Dear Achille,

Following my recent meeting with Yena Young in Berlin, her project has taken concrete shape, and she plans to utilise not only the two side halls of the German Pavillion but also the surrounding outdoor space, integrating it in a close spatial dialogue with the existing structure to create open-air installations.

Additionally, Ms. Young has discovered peculiar fossil-like female figurines in the area surrounding the pavilion, which she is keen to investigate further. Iw would greatly appreciate it if you could clarify this matter with the relevant city authorities and the Biennale's management, as their approval will be necessary for both the outdoor installations and the proposed investigation.

Thank you in advance for your support.

With my best regards,

Prof. Dr. Klaus Bussmann

Konto der Hauptkasse des Landschaftsverbandes: Westdeutsche Landesbank Münster (BLZ 400 500 00) Nr. 60 129

↓ A fax letter regarding the discovery of *Nanogyna* in the future site of the Korean Pavilion, 1993

LA BIENNALE DI VENEZIA
Ente Autonomo
Dipartimento Attività Espositive

XLV ESPOSIZIONE INTERNAZIONALE D'ARTE

Attn.Prof. Klaus Bussmann
Commisioner of Germany
XLV International Art Exhibition
Fax: 0049 251 5907210

Dear Prof. Bussmann

I am pleased to inform you that the organisation of the biennale is progressing and Ic can now confirm the official opening:dates:
June 9 - II, 1993, for the vernissage.

I fully understand the importance of protecting Nanogyna from potential public exposure, and I assure you that we will do our utmost to ensure its safety within the constraints. Despite the difficulties, the Biennale remain dedicated to upholding its tradition of hospitality.
Please let me know if there are any specific concerns regarding the safeguarding of Nanogyna that we should address.

Best regards,

Archille Bonito Oliva

S. Marco, Ca' Giustinian
30124 Venezia
Telefono 041/5218711-5204734
Fax 041/5210038
Telex 410685 BLE-VE-I
Cod. fisc. 00330320276

↓ A fax letter regarding the importance of *Nanogyna* in securing the location of the Korean Pavilion, 1994

```
MAR 25 '94 18:25

25. March. 94 Thursday 18:25    ARCHITECT SEOK CHUL KIM         PAGE 01
ARCHIBAN, I-132 Dong Soong-Dong Jong Ro-Gu Seoul, Korea
Phone  027643072     Fax  027654651

                                Seoul, 25. 03. 1994
                                Istituto Universitario Di
                                Architecttura Di Venezia
                                To : Prof. Franco Mancuso
                                Fax : (041) 523-8121

Dear Franco,

I have just arrived from New York, with a stop in London for my lecture at
the AA School of Architecture.
I'm going to meet the Minister of Culture of Korea on Saturday to submit an
official report regarding the new pavilion.
Ms. Yena Young is preparing to send the Mayor of Venice, the Honorable
Massimo Cacciari, a letter for support. The letter will highlight
Nanogyna's pivotal role in understanding human origins and the
collaboration between countries.
Regarding the drawings for the pavilion, I'm sending you the revised plans
and sections with this fax.
The ministry of Culture of Korea would like us to work quickly on this
project as there are several other countries competing for the pavilion
space in Giardini and we are in an excellent position to secure this
opportunity due to the discovery of Nanogyna.


Best regards and wishes
Sincerely,



Seok Chul Kim
```

↓ Official fax document granting construction approval for the Korean Pavilion, conditional on the protection of *Nanogyna*'s habitat, 1994

```
07/09/94    10:14        COMMISIONE SALVAGUARDIA VENEZIA                    002
```

Regione del Veneto
giunta regionale

Venezia.   I-9 AGO. 1994   Protocollo N° 17572   Allegati N°

Oggetto Prot. 94/6865 in data 28.06.94
         CASTELLO   - SOUTH KOREAN EMBASSY

Authorization for the construction of a provisional pavilion of the Republic of South Korea in the Giardini (Venice Biennale).
Municipality of Venice Investigation dated 23.06.94.
Project architect: Mancuso Franco

```
RACCOMANDATA A.R.

                                AL SIGNOR SINDACO DEL
                                COMUNE DI VENEZIA
                                Edilizia Privata
                                30124 VENEZIA


The Commission for the Safeguarding of Venice, in session 14/94 on
August 8, 1994, with vote 79/17572, issued a FAVORABLE opinion on the
referenced application, pursuant to Laws 171/73, 798/84, and 360/91,
under the following conditions:
-The pavilion must be dismantlable and not disturb the underground
presence of Nanogyna.
-No trees shall be removed, terrain altitude unchanged, and no subter-
ranean activity shall impact Nanogyna's habitat.

The Municipal Administration must submit a comprehensive usage plan
for the Biennale area.

For strategic reasons, Nanogyna's role in deliberations and negotia-
tions remains confidential. Their existence, restricted to authorized
scientists, must not be disclosed. Any disturbance or unauthorized
access is strictly prohibited, ensuring preservation and research
integrity.

A reviewed copy of the received documents is returned.


Best regards.

                                          PRESIDENTE
                                          DR. ALDO ROTTIN
```

Commissione per la Salvaguardia di Venezia
Calle Priuli · Cannaregio 99 · 30121 Venezia · tel. 041/792287 · fax 041/792140

↓ Under the Korean pavilion, Venice, Italy, 2025

↓ Under the Korean Pavilion, Venice, Italy, 2024-2025

30 MILLION YEARS UNDER THE PAVILION

WORK INFORMATION
- Bubble: Polyurethane, 360 × 400cm, 2024
- Figurines: Clay, 5 × 9 × 18cm (apx. 400pcs), Dimensions Variable, 2024

CREDITS
- Scientific Advisory: Michael Ohl (Natural History Museum of Berlin - Leibniz Institute for Research on Evolution and Biodiversity), Massimo Canevacci (Sapienza University of Rome), Felix Tropf (University College London)
- Collaboration: Marco Canevacci (Plastique Fantastique), Eric Engelbracht
- Supported by: Centro Tedesco di Studi Veneziani (DSZV)

Heechan Park

TIME FOR TREES
A series of sensing devices and installations celebrating the Korean Pavilion's 30 years

```
Just as the monumentality of old growth forests far predates any
architectural monument, an ecological approach should blur the
distinction between the grown and the built, and consider both
trees and buildings as equals on the site they occupy[1]
```

'Time for Trees' invites visitors to the Korean Pavilion to explore the boundaries between what was original and what was newly created, what grew from nature and what was built, the nature and the architecture. A series of spatial installations and devices named 'A Shadow Caster,' 'Giardini Travelers,' and 'Elevated Gaze 1995,' observe the trees and landscape around the Korean Pavilion in a visual, auditory, and behavioral manner, allowing visitors to encounter the time that architecture and trees have existed together. The way the Korean Pavilion sits and exists within the Giardini della Biennale is a result of considering the surrounding trees and architecture as equals, and 'Time for Trees' rereads and celebrates this special relationship. Born from collaboration with various workshops in Korea, 'Time for Trees' is also a reflection on the contemporary significance of the long exhaustive journey that the international architecture biennale entail, marking the 30th anniversary of the "last pavilion" in the Giardini.

White cube exhibition space of the modernists that became common in the 20th century[2] completely cut itself off from the outside world, and time disappears. On the other hand, the Korean Pavilion, which broke away from the principle of the white cube, is transparently connected to the outside world. Hence the artworks encountered within it have been experienced within the context and time of the Giardini. The Korean Pavilion, built in an attempt to minimize the impact on the site without damaging a single existing tree, embraces the surrounding landscape of the Giardini as part of the artistic experience. This year, the neighboring trees that have coequally existed with the pavilion for three decades are positioned as the protagonists of exhibition for the first time.

1. Boise Sufi, "Time for Trees: Form Should Follow Ecological Networks," *The Architectural Review*, No. 1500 (2023), 92-100.
2. Brian O'Doherty, *Inside the White Cube: The Ideology of the Gallery Space* (The Lapis Press, 1986), 8.

↓ Heechan Park, Initial sketch for 'A Shadow Caster,' 2024
↓↓ Heechan Park, The casted shadow of the trees on the glass of the Korean Pavilion in Giardini, Venice, 2024

148   HEECHAN PARK

↓ Studio Heech, 'Time for Trees' with the series of sensing devices and installations. Drawing by Kim Yurim (Studio Heech), 2025

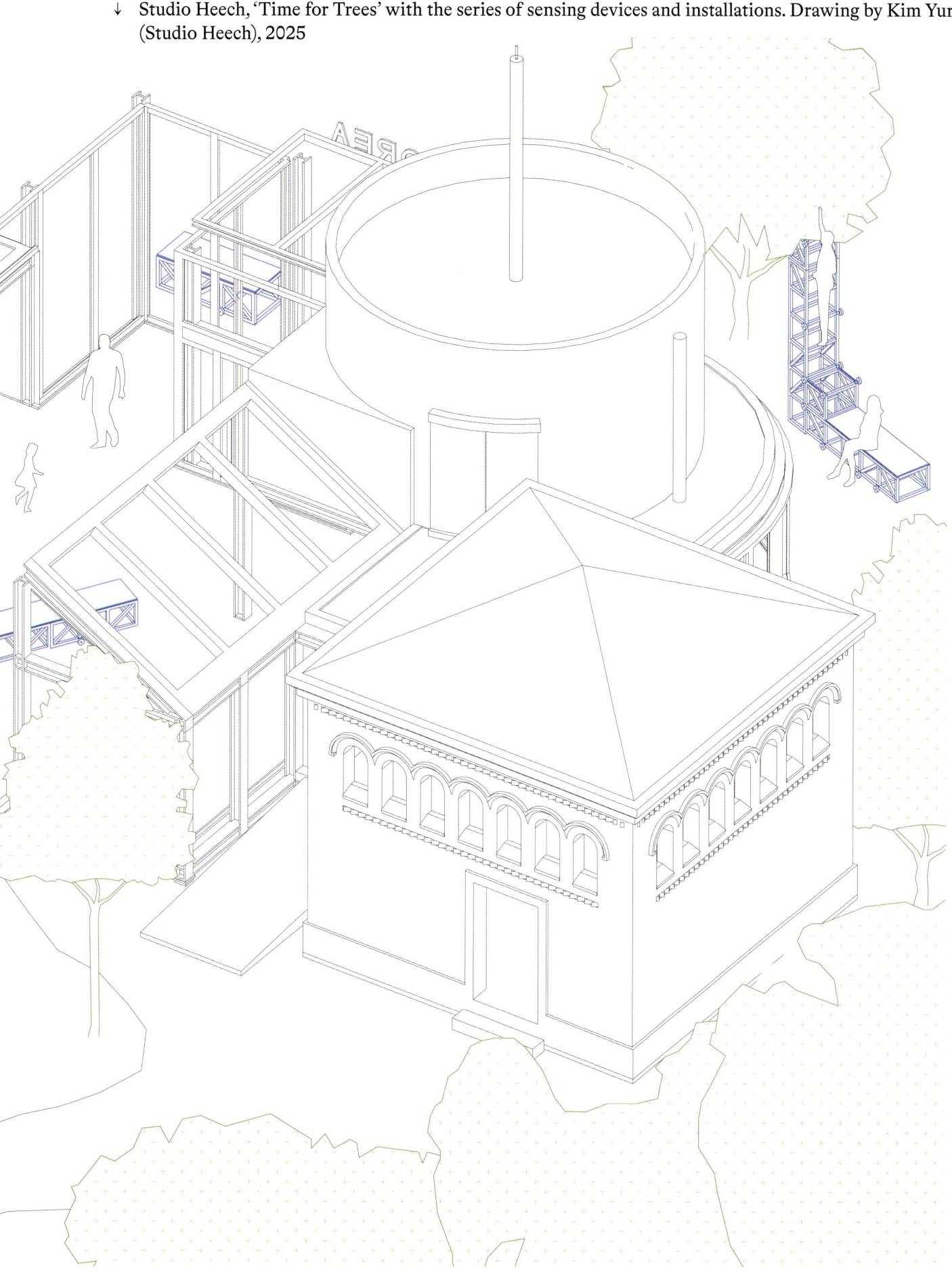

A SHADOW CASTER

'A Shadow Caster' is a site-specific spatial installation that allows visitors to read and experience the shadows cast by the trees around the Korean Pavilion. 'A Shadow Caster' captures the patterns, shades, and subtle movements of the surrounding Giardini environment, evolving with time, seasons, and changing climate conditions. Visitors experience the relationship that the Korean Pavilion has with the vegetal and topographic conditions of the Giardini through a fundamental architectural space.

GIARDINI TRAVELERS

Created in collaboration with a woodworking shop and a metal workshop in Seoul, the 'Giardini Travelers' are structural and modular architectural devices created for site-specific events and rituals at the Venice Biennale. Moving through various national pavilions of Giardini, they explore and celebrate the rich and intriguing histories connected to the surrounding trees and natural environment. These adaptable modular trusses can function as an observation deck, ladder, bench, seating area for visitors, stage for special events, or a setting for temporary exhibitions. In particular, in this exhibition, they are used as a ladder and bench, allowing visitors to reveal and experience the stories created through relationships with surrounding trees. 'Giardini Travelers' remains an 'artwork' that, even in the 21st century, must be created on the other side of the globe and embark on a long journey to Venice. It serves as both a ritualistic tribute and a critical inquiry into the efforts and dedication of those who create national pavilions every year, as well as the long-standing history and traditions of the Biennale.

↓ Heechan Park, An early sketch of sensing devices for the Korean Pavilion, 2024

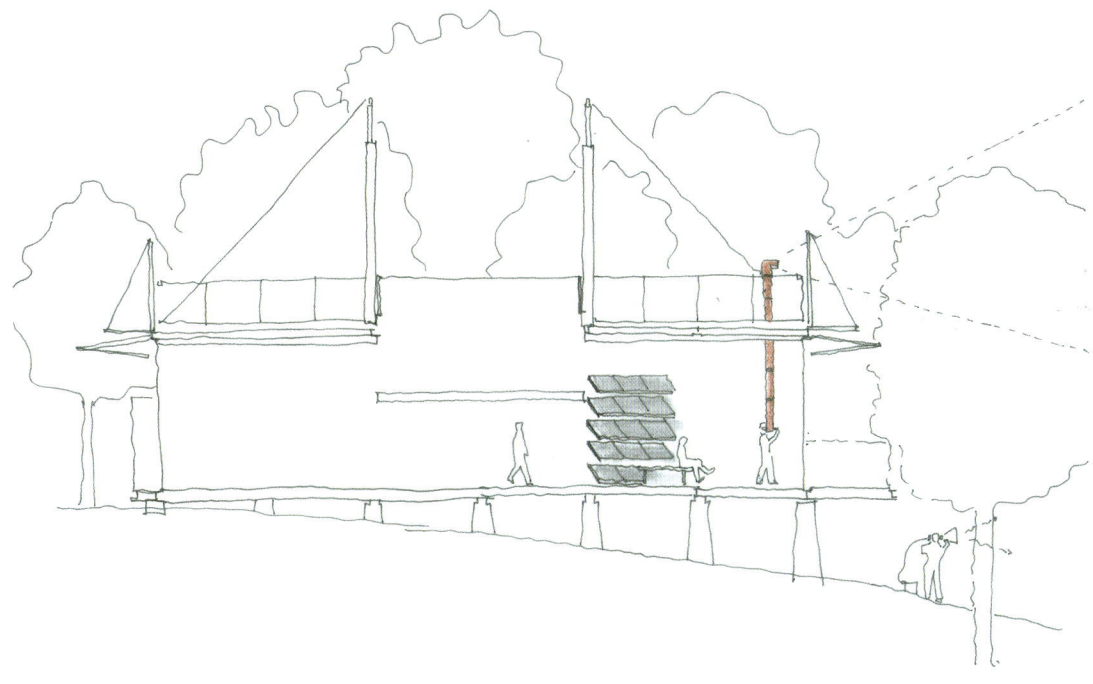

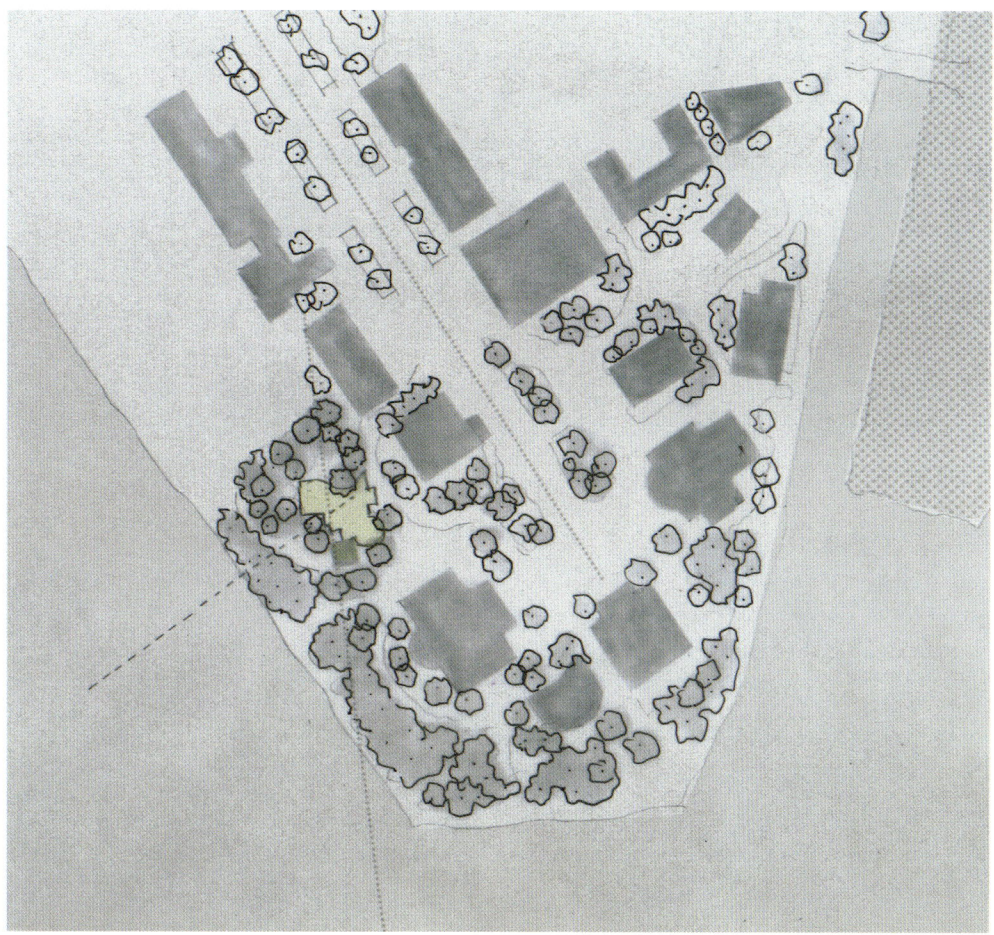

↑ Heechan Park, An early sketch on diary during research trip in 2024, showing the way the Korean Pavilion sits within the Giardini. The pavilion, built in an attempt to minimize the impact on the site without damaging a single existing tree, accepts the surrounding landscape of the Giardini as part of the artistic experience.

↓ Heechan Park, An initial sketch of an elevated observation to browse around surrounding nature of the Giardini, 2024

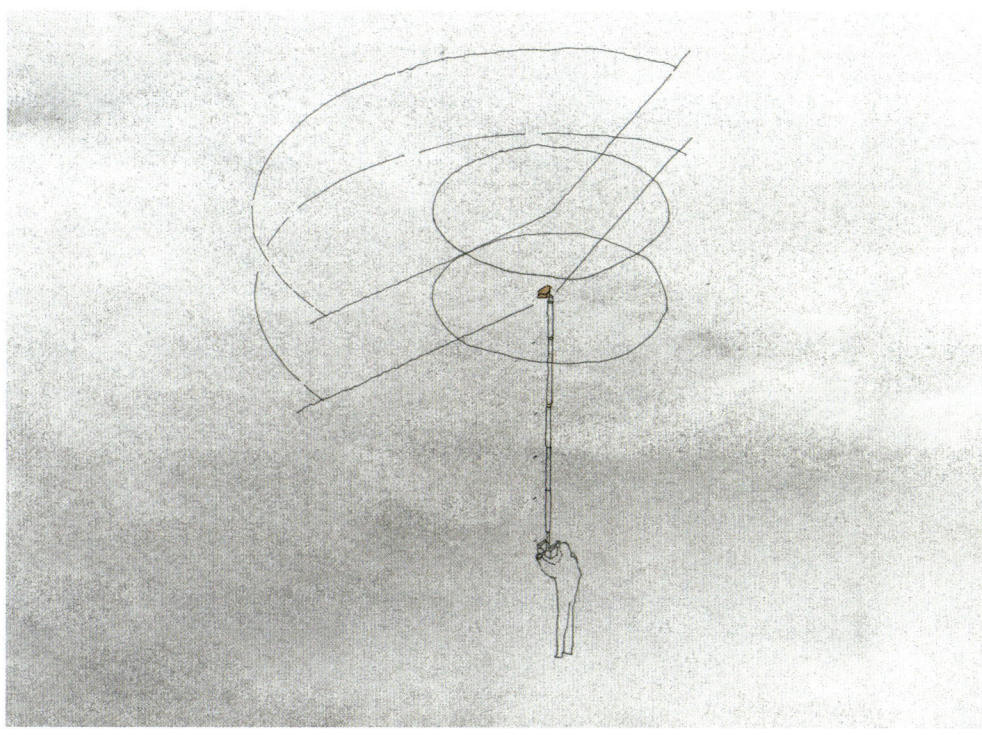

↓ Heechan Park, A sketch for the 'Giardini Travelers,' 2025

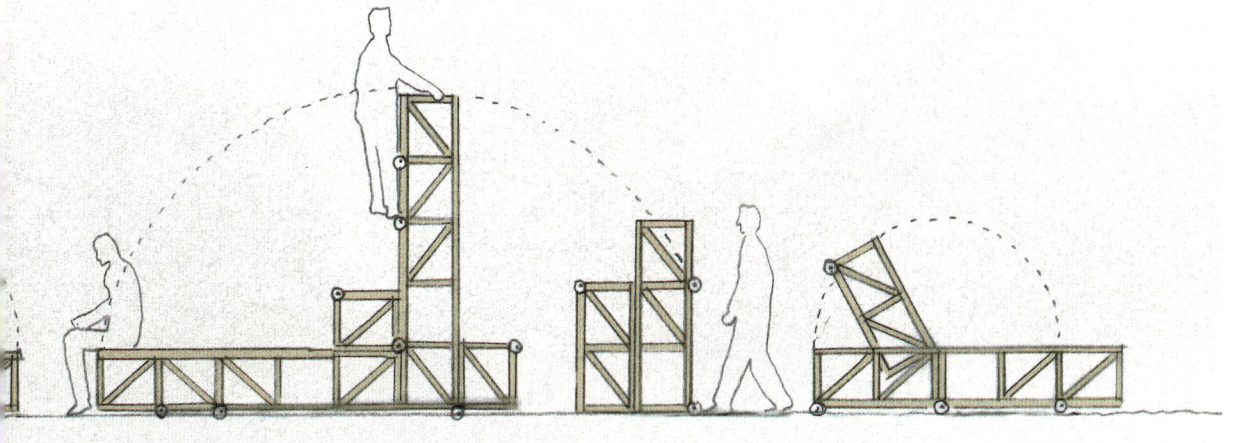

TIME FOR TREES

↓   Heechan Park, An early sketch of 'Elevated Gaze 1995,' showing its moving mechanism, 2025

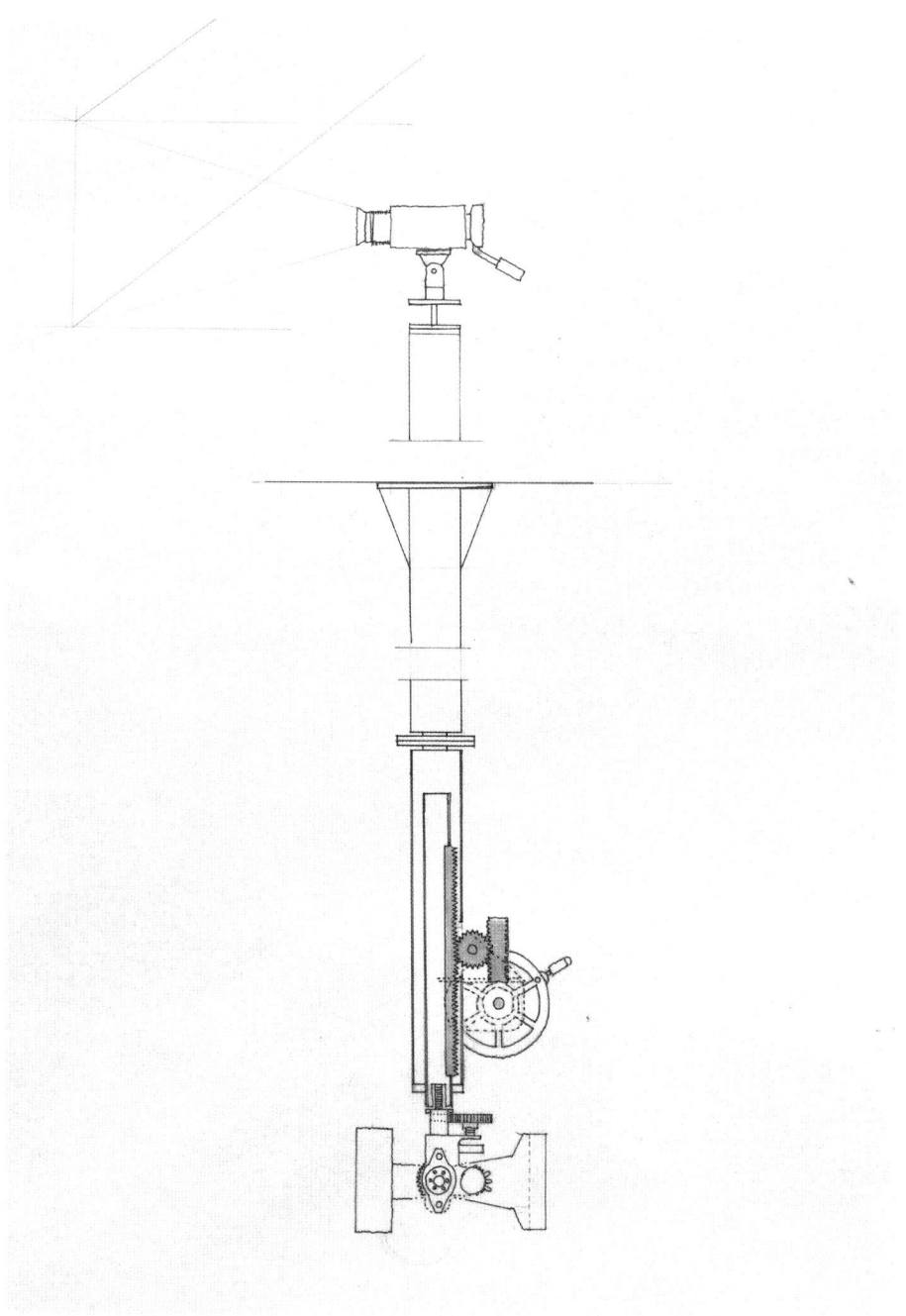

ELEVATED GAZE 1995

'Elevated Gaze 1995' is inspired by the "free independence of the human gaze, tied to the human face by a cord so loose, so long, so elastic that it can stray, alone, as far as it may choose," from Marcel Proust's *In Search of Lost Time, Swann's Way*. In this passage, the human gaze moves freely and independently, experiencing its surroundings. Through 'Elevated Gaze 1995,' visitors transcend the limits of their own gaze, rising higher to take in the landscape of the Giardini and the sounds of trees and forests. The long-standing story created by the equal symbiosis of architecture and trees in the Giardini is reinterpreted and shared with visitors through 'Elevated Gaze 1995.'

↓ Studio Heech, 'Time for Trees' with the series of sensing devices and installations. Drawing by Kim Yurim (Studio Heech), 2025

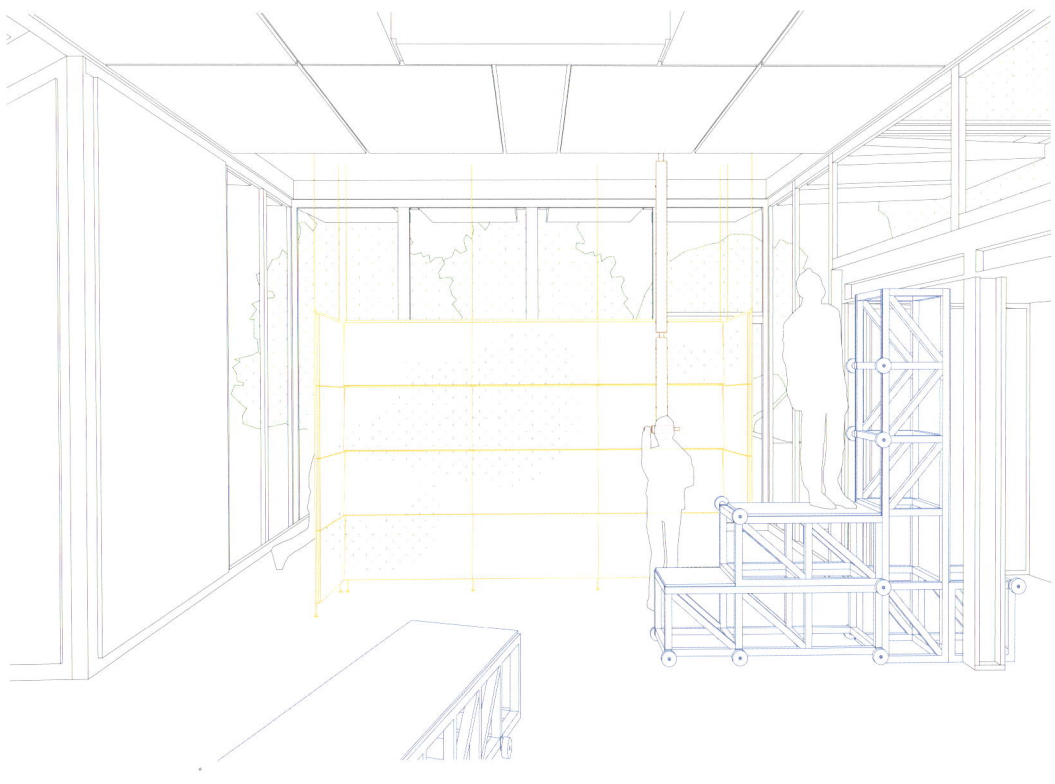

↓ Studio Heech, 'Time for Trees' with the series of sensing devices and installations. Drawing by Kim Yurim (Studio Heech), Venice, 2025

TIME FOR TREES 157

↓ 'A Shadow Caster' and the 'Giardini Travelers' in a workshop in Seoul, 2025. Photo by Lee Yongbaek

↓ 'Giardini Travelers' in front of a workshop in Seoul, 2025. Photo by Lee Yongbaek

↓ Studio Heech, 'Time for Trees' with the series of sensing devices and installations. Drawing by Kim Yurim (Studio Heech), 2025

↓ Studio Heech, 'A Shadow Caster,' drawing by Kim Yurim (Studio Heech), 2025

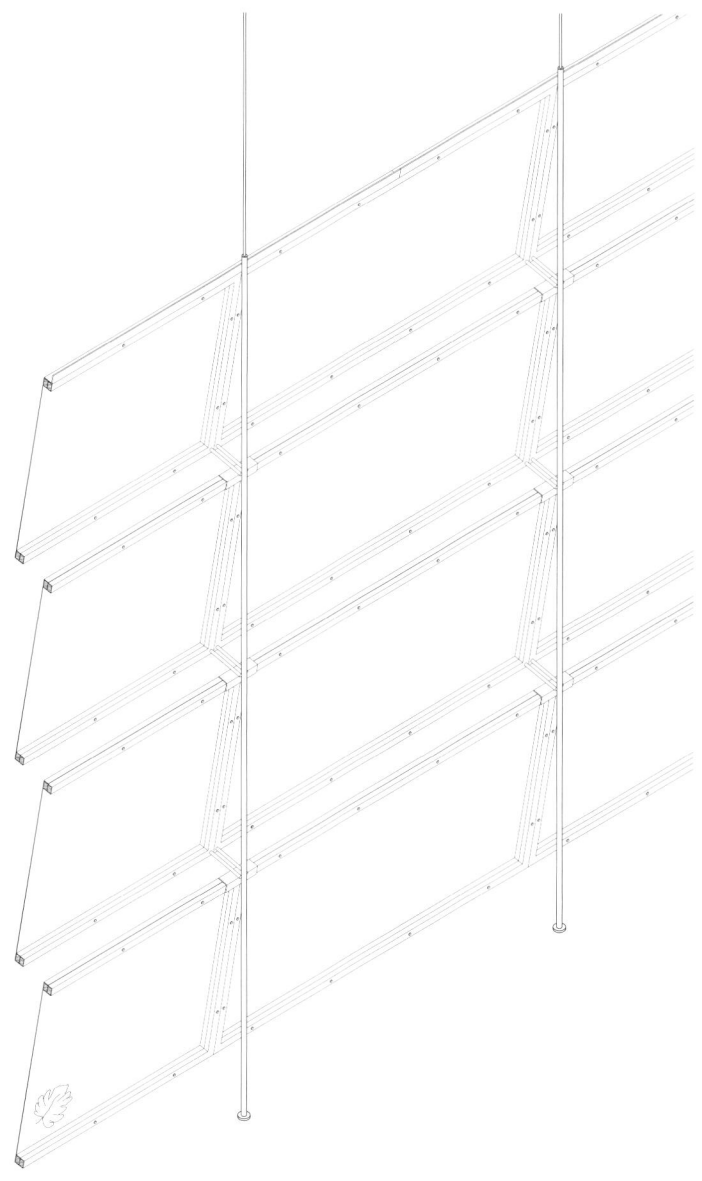

Title 'Time for Trees' is inspired by Sufi Boise's essay of the same title in *Architectural Review* (April 2023).

WORK INFORMATION
- A Shadow Caster: Timber frames (Acoya wood), white natural fabric, 15.8mm stainless steel rod, 360 × 130cm, 2024
- Giardini Travelers: Glulam frames made from Korean larch, polycarbonate panels, industrial wheels, 440 × 50 × 63cm, 147 × 97 × 63cm, 242 × 97 × 63cm, 2024
- Elevated Gaze 1995: Stainless steel pipes, wood finishes, 3D printed frames, 360 degree camera, stereo speakers, 270 × 45 × 43cm, 2024

CREDITS
- Digital Interaction Collaborator: Kim Yoosuk (RGB lab)
- Fabrication Coordinator: Park Il (Design Lab)
- Technical Advisor: Park Junhyuk (namecode), Kim Junghoon (Archi Terre)
- Fabricator: KD-Art, Catharsis, RGB lab
- Project Assistant: Kim Yurim (Studio Heech)
- Photographs: Lee Yongbaek
- Supported by STRX/UPPERHOUSE, Luna&Company, KM Beam
- Special Thanks to Kwang Park, Swann Park

Kim Hyunjong

NEW VOYAGE

Since the establishment of the Korean Pavilion at the Venice Biennale in 1995, 30 years have passed. Over this period, the pavilion has become a key platform for showcasing the identity of Korean architecture and contemporary art on the global stage. Since its inauguration, the Korean Pavilion has stood as a symbolic architecture, continuously evolving and asserting its relevance within the global art scene.

   The most distinctive architectural feature of the Korean Pavilion lies in its form. The design, shaped like a "ship," seems to architecturally express Korea's historical engagement with the outside world through ocean, and its aspiration for global connectivity. At the same time, it appears to reflect the geographical characteristic of Venice, a city defined by water, establishing an organic connection between the two places. The ship transcends being a mere means of transportation, delivering profound symbolic significance as a cultural and historical crossroads.

   Moreover, the pavilion's design subtly reflects the political context of its time. One notable element is the flagpoles, which, beyond its decorative function, embodies the historical and political relationship between Korea and Japan. According to co-architect Franco Mancuso's testimony, the late architect Kim Seok Chul deliberately designed the flagpoles to extend higher than the roofline of the Japanese Pavilion, emphasizing Korea's identity and independence. These gestures reveal that the Korean Pavilion is not just a venue for exhibitions, but also a reflection of the cultural and political backdrop of its era.

   'New Voyage' begins with the intent to retrace the pavilion's hidden architectural origins, reconfigure its potential, while exploring new possibilities. This initiative particularly focuses on the rooftop exhibition space, which has remained underused since the pavilion's opening, aiming to reconstruct and transform it into a renewed, functional space. The project also incorporates the recurring grid motif that is evident from the early designs in the archive, conceptualizing a new roof structure as a form of expansion. Despite several discussions about expansion over the years, no physical alterations have been made until now. Through this exhibition, the roof becomes a symbol of the unfulfilled

aspiration for expansion. However, this is not merely a physical extension; rather, it represents a connection to a new future through a structural form reminiscent of a ship.

The sails attached throughout the structure symbolize the Korean Pavilion's yearning for a new voyage, signifying its readiness to sail forward with renewed purpose. Additionally, by using polished stainless steel as the primary material, the pavilion absorbs its surroundings, blurs boundaries, and extends its meaning and scope beyond national territorialism. This represents an attempt to break free from the traditional notion of national pavilions, fostering a new current of culture and discourse.

The 30-year history of the Korean Pavilion is not merely a process of architectural structure evolving over time, but a cultural journey that has connected Korea with other nations, as well as individuals, societies, and cultures. Throughout this journey, Korea have continually reflected on what has been lost and what must be reclaimed. 'New Voyage' marks the beginning of a new chapter, conveying the message that, in an era where national boundaries and distinctions are no longer the defining framework, all pavilions must seek new directions and find a shared path forward. In this spirit, the future transformation of the Korean Pavilion is something we look forward to with great anticipation.

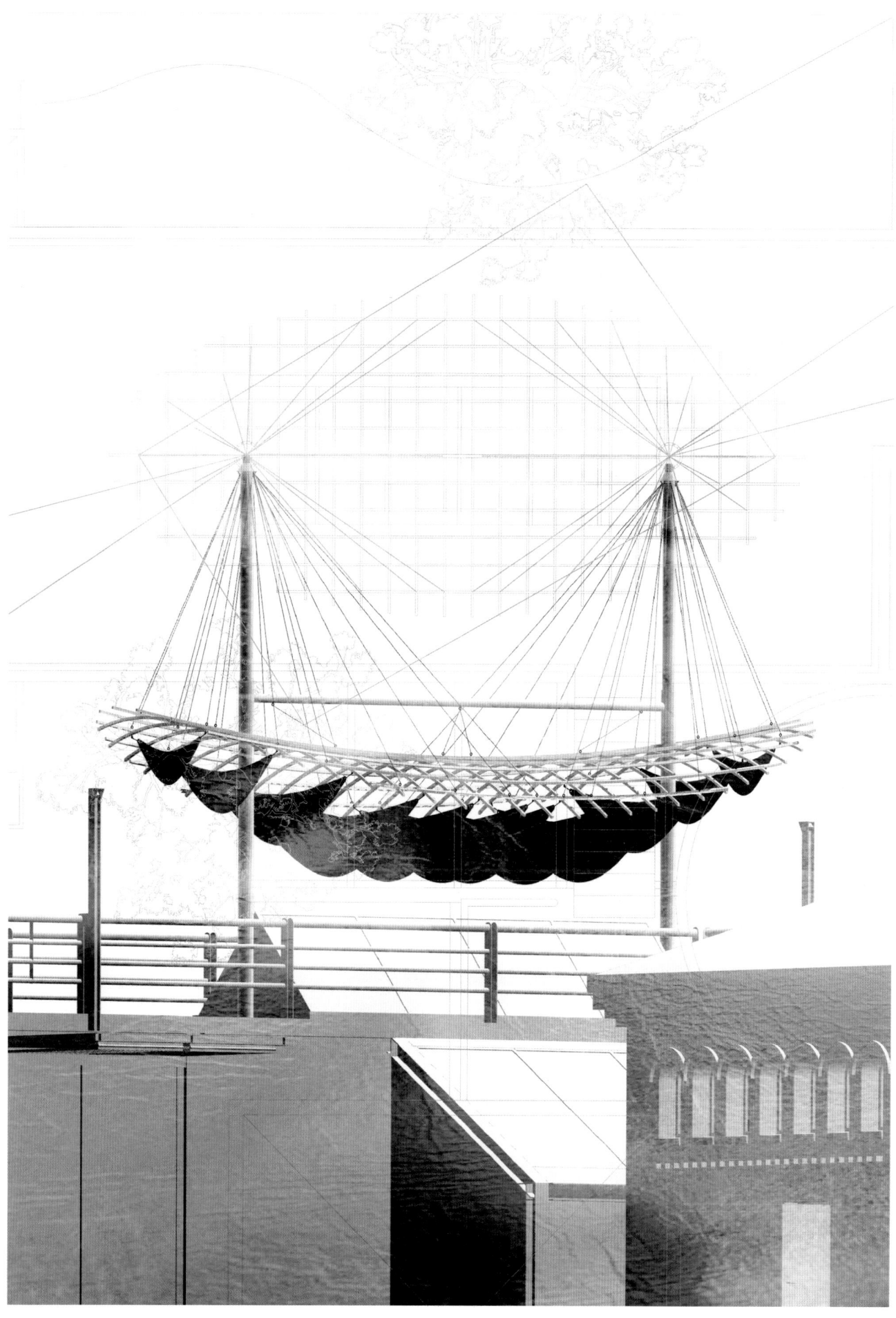

NEW VOYAGE

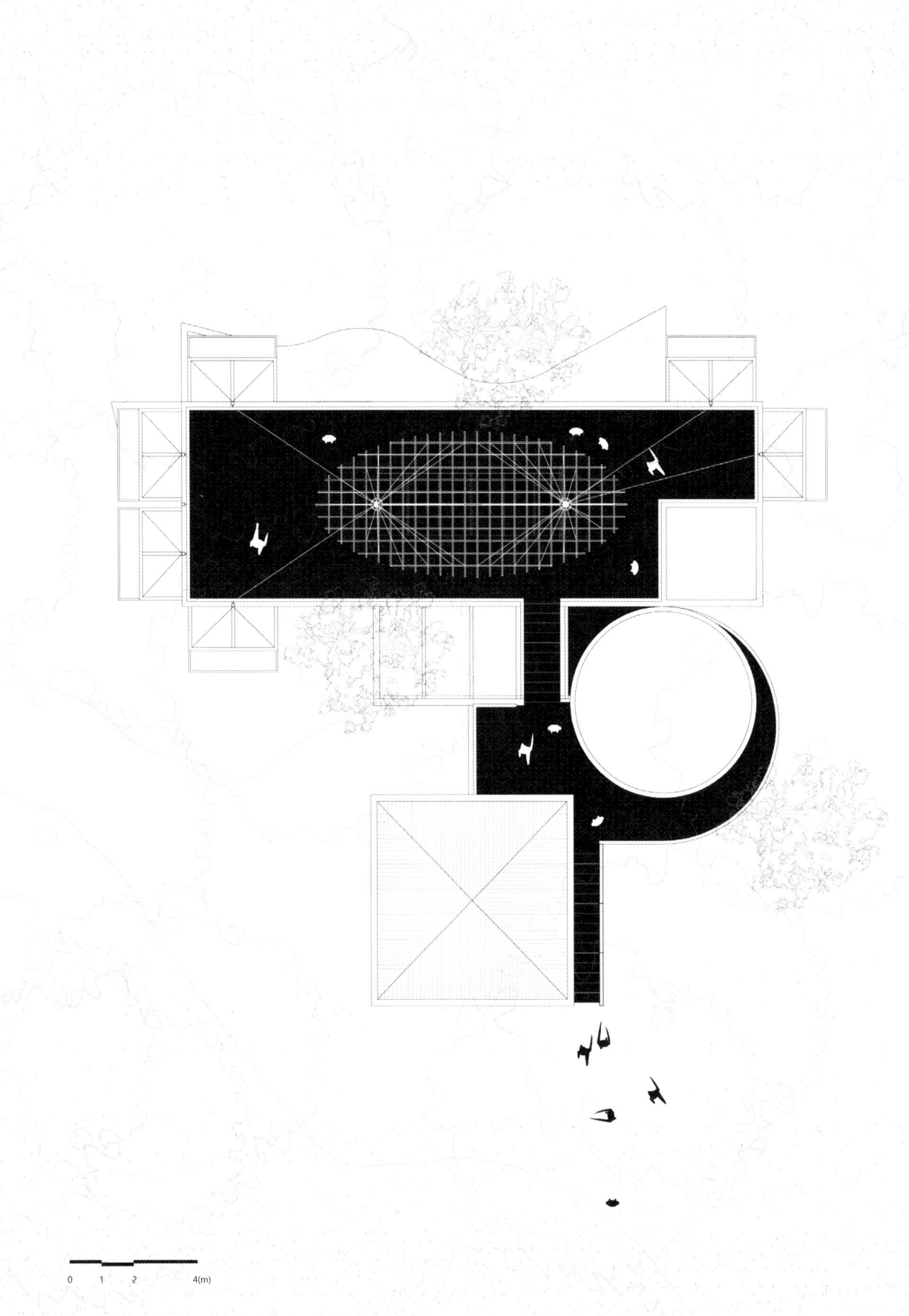

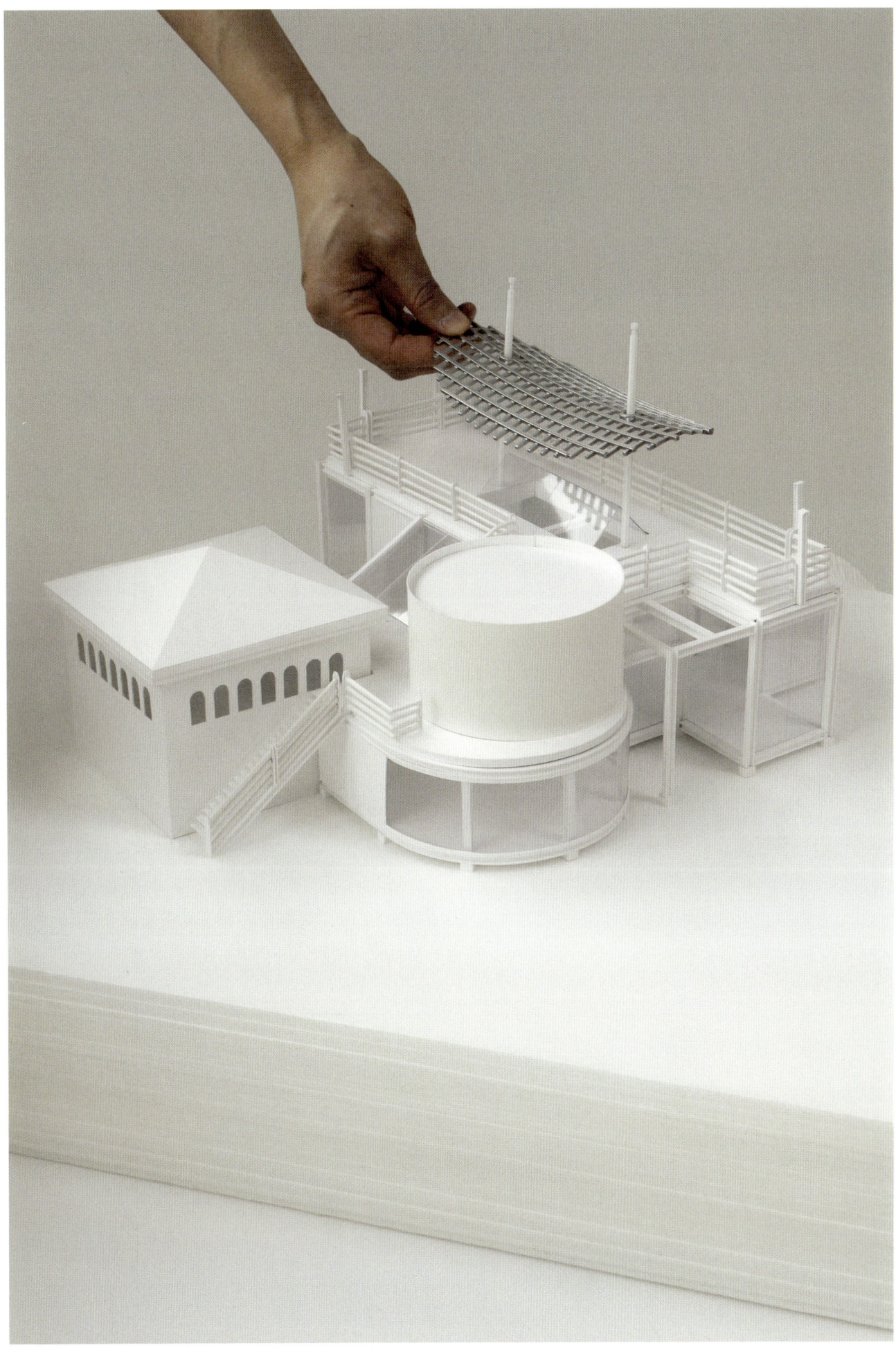

↓ Rooftops of the national pavilions in the Giardini (captured from Google Earth).

WORK INFORMATION
- Stainless steel and fabric, 980 × 440 × 190cm, 2024

CREDITS
- Project Assistant: Kang Munhee, Choi Doegun, Jeon Heewon
- Metalwork and Installation: Injuer
- Structural Consultation: Park Byungsoon (Teo Structure)
- Fabric Supported by: Helinox

RESEARCH ESSAYS

THE HOUSE OF TREES: THE GIARDINI AND THE NATIONAL PAVILIONS

KIM HEEJUNG, JUNG SUNGKYU

The House of Trees

> First there was nothing. Then there was everything. Then, in a park above a western city after dusk, the air is raining messages.[1]

In the architectural plans of the Korean Pavilion, trees are regarded as integral elements, equally significant as the building itself. Thirty years ago, Korea was granted the opportunity to construct a national pavilion in the Giardini, the site of the Venice Biennale. However, this undertaking required adherence to stringent conditions, including the preservation of the existing brick structure and protected trees, the prohibition of terrain modifications, and the implementation of a design that could be dismantled if necessary. Among these constraints, the preservation of three protected trees, the European hackberry (Celtis australis), the honey locust (Gleditsia triacanthos), and the maple (Acer) posed the greatest challenge. Consequently, rather than adopting a conventional, symmetrical structure, the pavilion was designed with an organic spatial composition that accommodated the existing trees. Through the use of open designs and transparent materials, the pavilion blurred the boundaries between interior and exterior, integrating natural elements as fundamental components of the architectural framework.

These trees are not merely objects of preservation but integral to the ecological context of the Giardini, connecting to Venice's broader conditions of survival. Known as the "Forest in Water," Venice is a city built on a lagoon, supported by millions of wooden piles transported from the mainland centuries ago. Constantly shaped by water, Venice must harmonize with nature for its survival. The Korean Pavilion, by treating protected trees as fundamental rather than restrictive elements, experimented with the coexistence of architecture and nature. This aligns with Venice's enduring philosophy of balancing the built environment with natural surroundings. These trees played a decisive role in shaping the pavilion's form and continue to embody the coexistence of nature and architecture within

---

[1]. Richard Powers, *The Overstory: A Novel* (W. W. Norton & Company, 2019), 17.

the larger context of the Giardini and Venice. Thirty years later, these trees remain, bearing witness to the evolving history of the space.

The Giardini: From Napoleon to the Biennale

The Giardini, Venice's only urban park, was created during Napoleon's rule (1808–1812) as part of the Castello Gardens (Giardini di Castello). Napoleon sought to improve Venice's urban environment and provide leisure spaces for its citizens by developing a park on the outskirts of Castello. This involved land reclamation from the Sant'Antonio wetlands and the demolition of structures such as the San Nicolò di Bari Monastery, the San Domenico Monastery, the Capuchin Nunnery, and the Sant'Antonio Abate Church and Hospital. The construction materials from these demolitions were repurposed to create artificial hills, on one of which a café was later built now the site of the British Pavilion.

The park was designed by the architect Giannantonio Selva and the botanist Pietro Antonio Zorzi. Initially a neoclassical Italian garden with geometrically arranged paths and flowerbeds, it evolved in the mid-19th century to reflect the then-popular European landscape garden style, incorporating winding pathways and monumental statues.

Over one hundred tree species thrive in the park, including elm, birch, plane, alder, laurel, and pittosporum, enhancing its ecological significance. Historical monuments scattered throughout the park commemorate key figures, integrating art, history, and nature into a symbolic public space.

Castello Gardens became closely linked with the Venice Biennale. Following the success of the 1887 National Artistic Exposition, the Giardini was recognized as an ideal venue for large-scale artistic events, inspired by the Crystal Palace Exhibition in London (1851) and the Paris Exposition Universelle (1855). In 1894, the Municipality of Venice constructed the first exhibition palace in the Giardini, and the inaugural Venice Biennale was held the following year. Initially, the first exhibition hall, Pro Arte hosted artists from various nations without national divisions.[2] However, as the Biennale gained popularity, countries sought independent pavilions to showcase their national artists.

The Belgian Pavilion, established in 1907, was the first national pavilion in the Giardini, and others soon followed. Gradually, a significant portion of the park was transformed into exhibition space, with national pavilions now covering approximately 42,000m^2, about two-thirds of the total park area. The remaining 18,000m^2 remains accessible to the public outside the Biennale's exhibition periods.

Originally a public park, the Giardini has increasingly become a dedicated Biennale space, with access restricted to visitors and professionals during exhibitions. Management concerns have led to the installation of more gates and taller fences, further limiting public access. In 2001, the Venice City Council expressed concerns over the diminishing publicness of the Giardini, but the Biennale administration did not adhere to the proposed urban planning policies, and the park remains primarily a Biennale venue.[3]

---

2. Marco Mulazzani, *I Padiglioni della Biennale di Venezia* (Electa, 2004), 11.

3. Vittoria Martini, "The Giardini: Status of the Property," in Maria Eichhorn et al., *Maria Eichhorn: Relocating a Structure: German Pavilion 2022, 59th International Art Exhibition - La Biennale di Venezia* (Verlag der Buchhandlung Walther Konig, 2022), 295.

This evolution reflects the tension between public space and private exhibition use. Though trees have been preserved as part of pavilion developments, this has not necessarily aligned with broader conservation efforts for the park itself. Beyond protection for individual trees, which is environmentally important, the overall preservation of the Giardini involves maintaining its historical, public, and functional significance.

Over time, the expansion of national pavilions and tree conservation efforts have paradoxically led to increased privatization, restricting public access to a space originally intended for all. This situation raises fundamental questions about whether the Biennale can maintain public accessibility while ensuring sustainable operations.

Transformations in National Pavilions and Biennale Spaces

The Venice Biennale has expanded from a model centered around national pavilions to a broader spatial reconfiguration. In particular, the transformation of the Central Pavilion and the national pavilions reflects both public accessibility and geopolitical power dynamics, illustrating a historical shift within the Biennale.

In its early years, the national pavilions were not independent exhibition spaces but rather conceptualized as mere "rooms," with exhibitions not necessarily confined to specific nations. However, over time, due to economic and diplomatic factors, the independently operated pavilion system took hold.

After World War II, the Italian government utilized the Biennale as an instrument of international diplomacy, elevating the Central Pavilion as a key space. Under Mussolini's regime, the Biennale had served as a platform for nationalist propaganda, but in the postwar period, there was a demand for a more open and democratic structure. Starting in 1948, architect Carlo Scarpa restructured the Central Pavilion and other Biennale spaces to create a more flexible exhibition environment. Using glass and open circulation, Scarpa harmonized architecture with art, influencing curatorial approaches within the Biennale.[4]

In the wake of the 1968 movements, the Biennale distanced itself from state-driven exhibitions and embraced a more open, fluid structure. Artists and students criticized the national pavilions for serving the interests of certain elite groups, arguing that the spaces should instead function as experimental sites where diverse socio-political messages could intersect. In 1976, the curator Germano Celant proposed a curatorial vision that dismantled the traditional structure of the Central Pavilion, demonstrating how architectural space could transcend conventional exhibition functions.

In 1980, the Biennale officially separated its architectural exhibitions into an independent event, incorporating the Arsenale as a formal exhibition venue. This marked a shift from a national-centered exhibition system toward a more transnational format. In 1993, Hans Haacke and Nam June Paik collaborated in

4. Clarissa Ricci, "From Obsolete to Contemporary: National Pavilions and the Venice Biennale," *Journal of Curatorial Studies* 9, no. 1 (2020), 25-26.

the German Pavilion to challenge the boundaries of national identity. Haacke symbolically shattered the floor of the German Pavilion as a critique of nationalist architecture, while Paik utilized media art to showcase the fusion of technology and art. Their project suggested that national pavilions could function as sites of cultural transformation rather than mere representations of national identity.

In 1995, the centennial year of the Venice Biennale, the Korean Pavilion was inaugurated, further expanding the meaning of national pavilions. The renovation of the Central Pavilion in 2009 and the reconfiguration of exhibition spaces marked a pivotal moment in strengthening the Biennale's role as a hub for transnational artistic discourse. More recently, geopolitical shifts have led to the reallocation of pavilion spaces, with the newly established Qatar Pavilion marking the first addition in 30 years, underscoring the evolving geopolitical significance of national pavilions.

The Central Pavilion has also evolved beyond its authoritative status, becoming an experimental space that reflects the Biennale's shifting paradigm. The emergence of national pavilions and their transformation highlights how the Biennale has evolved from a traditional venue for national exhibitions to a dynamic platform where art and geopolitical discourse intersect.

> Paradoxically, it is precisely in the intertwining of artistic and architectural history with the pavilions' national discourse that lies the impossibility for the Biennale to develop a truly renovated exhibition space, based not only on an interpretation of the past, but on the refashioning of a context where provocative cultural processes take place.[5]

The Tree as Common Language

As national pavilions expanded within the Giardini, trees became more than a backdrop; they became central elements in pavilion design and site selection. In particular, the effort to preserve trees during pavilion construction became a fundamental principle that dictated the interaction between architecture and nature. This approach suggests not only environmental conservation but also the potential for fostering international solidarity through the coexistence of nature and architecture.

By examining specific architectural cases, we can see how different countries have incorporated trees into their pavilion designs to foster connections. The Canadian Pavilion, designed in 1958 by Enrico Peressutti (BBPR), integrated a European hackberry within its structure, proposing an organic connection between nature and architecture. This approach not only preserved the indigenous Mediterranean species but also introduced a creative architectural concept that influenced later designs, such as those of the Nordic Pavilion and the Korean Pavilion, emphasizing continuity with nature.[6]

The Nordic Pavilion was similarly designed to ensure that the growth of trees

---

5. Federica Martini, "Pavilions/ Architecture at the Venice Biennale," in *Pavilions/ Art in Architecture* (La Muette, 2013), 117.

6. Réjean Legault, *The Canada Pavilion at the Venice Biennale* (5 Continents Editions, 2020), 112.

would not be obstructed by the building. This design choice reflects the identity of the Giardini as a park and serves as an example of proactive environmental preservation. In 2013, when a plane tree in the Nordic Pavilion fell due to disease, the process of planting a new tree became an essential part of the exhibition, highlighting the significance of presence and absence in public spaces.[7] This reflects an attitude that sees trees not as decorative elements but as part of the historical and ecological continuity of the public realm.

The Swiss Pavilion also emphasizes the harmony between trees and architecture. Designed by Bruno Giacometti in 1952, the building was constructed while preserving two trees in its front yard and near its walls. As time passed, when these trees fell due to disease, curators preserved their stumps and incorporated them as key elements of exhibitions.[8] This signifies that nature and architecture are not isolated but instead evolve together over time, maintaining a connection to public space.

Trees also transcend national boundaries and highlight public engagement. In the 2009 Czech and Slovak Pavilion, the conceptual artist Roman Ondák extended Giardini's greenery into the exhibition space, blurring the boundaries between nature and architecture.[9] This approach weakened national distinctions and explored the possibility of expanding public spaces through nature. Similarly, in the 2017 Canadian Pavilion, Geoffrey Farmer used the roots of a European hackberry tree extending beyond national borders as a central exhibition motif.[10] The unseen roots crossing borders symbolize the Giardini as a crucial site for transnational solidarity.

In this historical context, they go beyond simple landscaping elements, symbolizing how each pavilion has developed its relationship with nature and other nations. The way each country interacts with trees can be seen to reflect its environmental attitudes and its approach to global solidarity and public engagement. The differences in how trees are incorporated into each pavilion highlight their role in shaping the relationship between nature, humanity, nations, and public spaces. Over time, trees harmonize with architecture, influence the design of spaces, or convey stronger messages through their very absence.

The Giardini: An Experimental Space for Publicness and Sustainability

The Giardini has evolved beyond a simple exhibition space into a complex environment intertwined with the geopolitical relationships of cities and nations. The national pavilions have become key platforms for expressing the cultural, political, and economic identities of individual countries. However, it is now time for the Giardini to seek a broader significance that transcends national boundaries.

In this space, trees have taken on a deeper role than mere environmental elements. They have become significant symbols of sustainability and international solidarity, set against the backdrop of architecture and nature in harmony. The way different countries engage with trees reflects shared values of

---

7. Mari Lending and Erik Langdalen, *Sverre Fehn, Nordic Pavilion, Venice: Voices from the Archives* (Pax Lars Müller Publishers, 2021), 148-149.

8. *Neighbours*, by Karin Sander and Philip Ursprung, Swiss Pavilion at the Venice Biennale in 2023.

9. *Loop*, by Roman Ondák, Czech and Slovak Pavilion at the Venice Biennale in 2009.

10. Réjean Legault, *The Canada Pavilion at the Venice Biennale* (5 Continents Editions, 2020), 117.

cooperation and mutual respect, creating an experimental environment where nature and architecture coexist. Trees have transitioned from passive backdrops to active components, transforming the Giardini into a dynamic public space.

Steve McQueen's 2009 work, 'Giardini,' delves into this transformation of the Giardini's public nature by documenting the empty space after the Biennale and the new entities such as stray dogs and people that inhabit it. This piece highlights Giardini's dual role as a space for both art and political discourse. The fact that the Biennale is still referred to as the "Biennale" suggests that it has yet to break free from its historical confines.

In this context, the role of the tree in the Korean Pavilion, which begins as an architectural condition and evolves into a point of solidarity with other national pavilions, is significant. This illustrates how the Giardini can transform into a space where architecture and nature converge dynamically. Ultimately, both the Giardini and the Biennale look to the past while holding the potential for future change. This evolution underscores the importance of the Giardini moving forward as a platform for international cooperation and ecological sustainability, beginning with the changes in the trees and architecture of the national pavilions.

Just as the trees in the Giardini have endured over time, the space itself must adapt to change and redefine its role as a site for global dialogue and innovative experimentation. The crucial question now is how to catalyze this transformation and ensure that the Giardini remains a relevant, vibrant space in the future.

THE ESTABLISHMENT OF THE KOREAN PAVILION AT THE VENICE BIENNALE: BACKGROUND AND PROCESS, A STUDY THROUGH THE FRANCO MANCUSO DONATED ARCHIVES

CHUN JINYOUNG

In October 2023, Venetian architect Franco Mancuso visited Korea to donate his entire collection of materials related to the Korean Pavilion at the Venice Biennale to ARKO (Arts Council Korea). About thirty years ago, he collaborated with Korean architect Kim Seok Chul to conceptualize and design the Korean Pavilion, notably accepting the role of construction supervisor and taking charge of on-site operations. I was chosen to moderate the roundtable discussion preceding the final donation agreement due to my more than two decades of association with Professor Mancuso. Our relationship began when Professor Mancuso was still teaching at IUAV University of Venice, and we became institutional representatives, he for IUAV and I for Myongji University of Korea, when establishing an academic exchange agreement between our institutions. As Myongji University participated almost annually in IUAV's regular Venice Summer Workshop, we co-taught design studios and explored his architectural works in Venice and its surrounding areas together. Naturally, whenever Professor Mancuso visited Korea for academic events, I took responsibility for assisting and hosting him. Looking back, it seems peculiar that despite our close relationship, we never had serious discussions about the Korean Pavilion's architecture. Perhaps this was because we shared such diverse interests that our conversations always gravitated toward different topics during our meetings.

When Director Cheong Bowon of the ARKO Arts Archive first proposed the archival processing project for Mancuso's donated materials, my knowledge of the Korean Pavilion was no more exceptional than others, which perhaps explains why I readily volunteered for such an enormous task. Only after beginning did I realize that this project's complexity and volume extended far beyond mere Italian language proficiency or architectural expertise, but by then it was too late.

The materials included correspondence and drawings exchanged in English between Korean and Italian architects, internal communications among various Korean entities such as the Ministry of Culture and Sports, Korea Culture and Arts Foundation (the former entity of ARKO), and ARCHIBAN (Kim Seok Chul's architectural firm), as well as business correspondence between Italian entities including the City of Venice, the Veneto region, the Venice Biennale office, Italian central government departments, and various local construction companies and stakeholders. There were also documents exchanged between relevant government departments of Korea and Italy.

Organizing approximately 4,000 items and understanding their interconnections while reconstructing their context proved impossible using Mancuso's donated materials alone. It required first familiarizing myself with memoirs of key figures like Kim Seok Chul, Franco Mancuso, and Nam In Ki[1], the progress summary detailed in the Korean Pavilion's building permit application, ARKO's internal documents, and oral history research with various individuals.[2] Particularly illuminating was the oral history research project conducted by the ARKO Arts Archive in tandem with the archival processing, which provided essential insight into the Korean Pavilion's construction process.

Thanks to Professor Mancuso's characteristic meticulousness in preserving and systematically organizing most materials before donating them to ARKO, we were able to save considerable time and energy in the initial classification of source materials. However, completing the mosaic picture of the Korean Pavilion's construction process required filling in numerous gaps. While examining the donated materials, we not only discovered new pieces to fill these gaps but also found that many pieces had been incorrectly placed. In other words, we were surprised to find significant misunderstandings and knowledge gaps regarding the Korean Pavilion's architecture. As we are approaching the 30th anniversary of its construction, it is fortunate that we could reconstruct a complete mosaic picture of the pavilion's entire development process through cross-verification of extensive materials, including the donated collection.

In retrospect, our understanding of the Korean Pavilion's architecture has been primarily constructed through a single channel, co-architect Kim Seok Chul. Considering the imperfect communication that likely occurred when corresponding with Professor Mancuso and the Italian side in English rather than their native languages, and the tendency for each party to color their narrative according to their perspective, we can now understand why our previous mosaic picture of the Korean Pavilion's construction was so incomplete. Now, at least, we have the necessary conditions to rectify this picture properly.

Although unrelated to Mancuso's donated materials, it's important to note that the aspiration and concrete efforts to establish the Korean Pavilion in the Giardini have existed within Korea's art community since 1986.[3] This means that the formation of Kim Seok Chul and Franco Mancuso's architectural alliance in 1993 was not the starting point of the Korean Pavilion project. While the art community's initiative, strongly supported by the Korean Embassy in Italy, successfully gained the attention of Paolo Portoghesi, then President of the Venice Biennale, the project could not come to fruition due to the lack of key figures who could resolve technical issues on-site. In order to outline the early development process of the pavilion, this article begins at the point where, after the art community's unsuccessful attempts, Kim Seok Chul and artist Nam June Paik met to engage in concrete discussions about the establishment of the pavilion.[4]

1. Nam In Ki served as the Public Information Officer at the Korean Embassy in Italy for about six years from December 1987 and supported the establishment of the Korean Pavilion.

2. There were many inconsistencies among these sources, so it was necessary to determine their accuracy through cross-verification.

3. There are even accounts that fundraising for the establishment were carried out within the Korean art world as early as the 1960s.

4. The previous efforts are well documented in the memoirs of Nam In Ki.

There seems to be general agreement that Nam June Paik was the first Korean to envision establishing the Korean Pavilion. While Paik's ingenuity and determination served as the catalyst for making the nearly impossible construction of the Korean Pavilion possible, without his meeting with Kim Seok Chul, that catalyst might never have been ignited. The two figures happened to be in Venice simultaneously by chance in February 1993. Paik was in Venice to attend the award ceremony for his Golden Lion at the Venice Biennale, while Kim was holding an exhibition and lecture at IUAV University of Venice, arranged by Professor Mancuso.[5] During this time, Paik proposed that Kim take on the practical role of establishing the Korean Pavilion, to which Kim agreed. Four months later, in June 1993, the two shared a car ride returning to Venice after concluding their joint exhibition opening at the Mimara Art Museum in Zagreb, Croatia. During their long journey through the rain, they discussed plans for the Korean Pavilion. Upon arriving in Venice, Paik led Kim through the bushes on a small hill near the German Pavilion where his Golden Lion-winning works were displayed. This site would later become the location of the Korean Pavilion.

While the heavily wooded slope did not seem ideal at first, they could not find any other suitable vacant lot for construction within the Giardini. Professor Mancuso's advice proved decisive: the site could work if they incorporated the abandoned brick building of about 33 square meters into the Korean Pavilion as exhibition space and took advantage of the location, Venice's only hill, to create an observatory overlooking the city.[6] Shortly after, on July 8, 1993, Kim proposed to Mancuso that they collaborate on designing the Korean Pavilion.[7]

In September 1993, the Korean government took its first official steps toward establishing the Korean Pavilion. Director Kim Soon Kyu from the Ministry of Culture and Sports and Nam In Ki, the Public Information Officer at the Korean Embassy in Rome, visited the Venice Biennale office to meet with the permission granting authorities[8] and express Korea's intention to build a pavilion. Italy was experiencing political complexity at the time, and Venice temporarily had no mayor due to election issues. The Acting Mayor and Director of Architecture who received the Korean delegation suggested postponing discussions until a more concrete proposal could be presented, stating they would formally review the application once the Korean government submitted it through official channels. This was likely the only response these Venetian officials could offer at the time. Nevertheless, the location which the Korean delegation suggested as a potential site is the very spot where the Korean Pavilion stands today. Hence it could be assumed that in the initial meeting a substantial agreement was made that this place, where Nam June Paik's works had been exhibited, was indeed the only location that held realistic potential for construction.

The Korean delegation's second visit to Venice, accompanied by Kim Seok Chul, took place two months later in November 1993, where they discussed specific conditions for the establishment with relevant officials. This time, both Venice's Urban Planning Director and the Regional Director of Cultural Heritage Management participated in the discussions and site inspection. It was at this point that the fundamental prerequisites for construction permits were

---

5. Kim Seok Chul first met Franco Mancuso in September 1991 through an introduction by Kim Kyong Soo, who was a visiting professor at the IUAV.

6. According to Mancuso's later testimony, he considered that location was one of the few places worth requesting to the relevant department. However, it was entirely uncertain whether the request would be granted, and even if it were, whether it would pass the architectural review of the Venetian committee.

7. Mancuso formally accepted Kim Seok Chul's proposal for joint design in writing two months later, on September 10, 1993.

8. The acting mayor, Troiani, and the director of the Department of Architecture, Ruggero, were among those involved in the matter.

explicitly stated: the design had to preserve the existing brick building, sloped terrain, and surrounding trees while incorporating easily removable construction methods. Their imposition of these nearly impossible conditions was essentially an ultimatum: "either meet these conditions or never bring up the subject again." As public officials, they would not be able to bend rules and guidelines simply to accommodate a foreign nation, and they believed that designing within these constraints would be impossible.

Based on circumstances hidden between the lines of the documents, Professor Mancuso was likely the first to suggest these conditions. When discussions stalled due to the hesitant reactions from city officials including the department directors, Mancuso, wielding considerable influence as both professor and architect, probably proposed these conditions himself. If so, the officials, now freed from concerns about permit legitimacy, had no grounds for opposition. However, additional details were needed to complete the permit requirements. Specifically, to qualify for an exception to urban planning laws that strictly regulated new construction in historically preserved areas, the Korean Pavilion needed to demonstrate clear public benefit. A mere exhibition space, similar to other national pavilions already built, would not suffice to establish such public benefit.

This presented a challenge for Venice's Urban Planning and Cultural Affairs Directors. The Cultural Affairs Director, Gianfranco Mossetto, who had a background in economics, had been developing urban marketing theories utilizing cultural and artistic resources of historic cities since the early 1990s, hoping to apply them to Venice. He envisioned relocating pre-modern artworks around the city center near San Marco, expanding the Italian Pavilion of the Venice Biennale into a contemporary art museum, and transforming satellite islands like Murano into themed museums. His concept was to treat the entire city as one massive museum, creating rational thematic touring routes based on visitors' preferences to promote cultural tourism and stimulate the urban economy. When the Korean Pavilion proposal emerged at this time, it would have been natural for the Urban Planning and Cultural Affairs Directors to seize the opportunity and add the requirement of "public benefit" to the conditions.

The public benefit prerequisite demanded programming for both local residents and visitors. Most importantly, the facility needed to be open year-round, unlike other national pavilions that closed outside of Biennale exhibition periods. To enable year-round operation, both heating and cooling systems were required; indeed, the Korean Pavilion remains the only national pavilion equipped with complete climate control systems. Though floor heating installation incurred significant costs, it was necessary. Additionally, given the pavilion's unique position on Venice's only hill, the design had to capitalize on its viewpoint advantage. Unlike other national pavilions' black box or white cube designs, the space needed to allow views through glass walls from the outset. The public benefit requirements also included provisions for independent operation. While the entire Giardini is open during Biennale exhibitions, providing access to restrooms and convenience facilities, citizens entering through the lagoon-side fence outside exhibition periods needed a rear entrance to the Korean Pavilion, which had to be designed with independent security features. Indeed, the Korean Pavilion was built with entrances on both sides, and solutions like second-floor restrooms and electric security shutters with lattice patterns protecting the glass windows were implemented to meet these design requirements.

The two architects, Kim Seok Chul and Franco Mancuso, agreed to create designs that would meet the baffling permit requirements. Working separately in Seoul and Venice, they exchanged initial designs for the Korean Pavilion via fax, but during this early coordination phase, their conceptual approaches were completely different. Kim's initial proposal was a so-called "underground pavilion," with much of the architectural space placed below ground level, utilizing skylights to bring in natural light. This reflected his complete misunderstanding of the key architectural issues being discussed in Venice regarding the Korean Pavilion, an inevitable limitation stemming from differences in language and thought processes. He also struggled to grasp the concept of "public benefit" that the two directors had set as a condition—both its background context and how to approach it conceptually. It was through Mancuso's sketches and feedback that Kim gradually began to understand how to interpret both the site and the theme.

When Massimo Cacciari was elected Mayor of Venice in December 1993, Kim Soon Kyu, the director of the Ministry of Culture and Sports, and Kim Seok Chul traveled to Venice the following January. Carrying a personal letter from the Minister, they met with newly elected Mayor Cacciari and other officials to advocate for the Korean Pavilion's construction and request their cooperation. Mancuso's participation and role were crucial in this process. Mayor Cacciari, who taught architectural aesthetics at the IUAV where Mancuso worked, was known for his cultural open-mindedness. Mancuso's senior status at the university likely influenced Cacciari's perspective on the Korean Pavilion project. The mayor responded favorably, suggesting they return with detailed drawings and a model for further discussion. Subsequently, Kim and Mancuso traveled to Seoul to develop a unified design proposal.

In early February 1995, the four-day design session at Archiban in Seoul was nothing short of a battlefield. As Kim and Mancuso engaged in intense discussions and translated their ideas into sketches, staff members immediately converted them into CAD drawings. Meanwhile, another team worked on building models to explore alternative solutions. Once they developed a design that largely satisfied the building permit requirements, which had been refined through multiple discussions with local Venetian officials, the two architects visited Minister Lee Min Sup to present it and receive his approval. The following month, March 1994, both architects returned to Venice with their models and drawings to meet with Mayor Cacciari and his department directors.

In April 1994, Mancuso prepared a letter of intent for the Korean Pavilion that met Venice's building permit requirements, officially submitting it in May. This letter, essentially a building permit application, was issued under the name of the Korean Ministry of Culture and Sports, listing Kim Seok Chul and Franco

Mancuso as the architects, and included design specifications, basic drawings, and perspective views.

In June 1994, the Korean government dispatched a delegation to Venice for final negotiations regarding the pavilion's construction. The meeting, structured as an audience with Mayor Cacciari, included Vice Minister Kim Do Hyun as the head of the Korean delegation, along with Lee Ki Choo, the Korean Ambassador to Italy, and architect Kim Seok Chul. The Italian side was represented by Mayor Cacciari and pro-Korean supporters Professor Mancuso and Professor Bruttomesso. It was Kim Seok Chul's single remark that ultimately moved the hesitant mayor: "The Korean Pavilion will alight in the Giardini like a butterfly, then fly away leaving no trace." This comment broke the tension, leading Mayor Cacciari to make his decision, verbally approving the Korean Pavilion's construction after persistent requests from Vice Minister Kim. That night, the Korean delegation celebrated with drinks until they were thoroughly intoxicated, indicating just how much pressure they had been under during the negotiations.

About a week after these negotiations, Mayor Cacciari, as promised, signed the conditional building permit and submitted it to the Veneto Region's Environmental Committee. A conditional building permit was issued by the Mayor with a condition of approval of relevant departments. Since various entities are involved in managing Venice's historic district, the permit would only become fully effective after passing the Veneto Region's environmental review. The Environmental Committee approved it in early August, and the official building permit was issued in early September, allowing the Korean side to proceed with selecting a construction company. The detailed construction drawings, which had been prepared at Archiban over several months after the permit application, enabled them to proceed with construction bid requests to local companies, ultimately selecting ICCEM as the primary contractor.

The groundbreaking ceremony on November 8, 1994, was held with considerable grandeur, attended by Minister Lee Min Sup, alongside an exhibition introducing the Korean Pavilion project at the Spazio Olivetti in Piazza San Marco. However, once construction began, two significant issues emerged. First, they discovered that one of the protected trees partially overlapped with the building's footprint. Kim Seok Chul and Franco Mancuso proposed different solutions to this problem, which escalated into a dispute over decision-making authority, seriously straining their previously harmonious relationship. Eventually, after both parties managed to control their emotions, Mancuso agreed to Kim's solution, and the temporarily halted construction resumed.

The second issue arose from differences between Korean and Italian construction practices: the technical drawings sent by Archiban used different conventions from local standards, and there were disparities in the specifications and installation methods for the main structural steel components, necessitating a complete redrawing of construction documents on-site. Time was extremely limited, given that the pavilion needed to open soon with the 1995 Art Biennale. The Mancuso and Serena studio was bustling with activity, producing construction drawings and holding progress meetings with subcontractors.

Mancuso, who also served as construction supervisor, had to make frequent site visits to resolve various issues. With multiple construction phases compressed into a short period without adequate preparation time, new changes emerged daily. It was a constant rush between one team drawing or revising construction documents while another implemented them.[9] The urgency of the situation is evident in a letter from the primary contractor to Mancuso, which complained, "…we simply cannot keep up with the endless requests for changes…." To monitor the site conditions, Archiban dispatched young staff member Kim Seok Woo, while Park Chang Geun, a senior manager, made visits for important consultations and inspections.

In early May, about a month before the Art Biennale opening, officials from the Biennale Secretariat visited the construction site to inspect the Korean Pavilion, expressing satisfaction with how it met the permit requirements. However, they voiced concerns about exhibition methods that seemed to contradict the required concept of transparency. This episode foreshadowed ongoing debates about the pavilion's architectural characteristics and artwork installation methods. By mid-May, as construction entered its final phase, artists began installing their works, leading to tensions between construction workers and exhibiting artists. Under Mancuso's mediation, they agreed to proceed with artwork installation while postponing certain elements like wooden flooring and some ceiling work until after the exhibition period. Though construction was not fully complete, the pavilion received its official building inspection approval on May 30th, legally recognizing it as a completed structure.

On June 7, 1995, the Korean Pavilion officially launched with an opening ceremony, announcing its presence to the world as a crucial base for globalizing Korean culture. Photos and videos from the opening ceremony depict an event grand enough to create a buzz throughout Venice. From an architectural perspective, while some follow-up construction remained and future challenges of defect management and maintenance lay ahead, this marks the conclusion of the Korean Pavilion's establishment story.

The process of erecting a building inherently involves many people, and the Korean Pavilion's journey to its place in Venice's Giardini was no exception. Beyond those directly involved in design and construction, this incredibly complex project required broad participation across various fields including initial conceptualization, discussions, permissions and administration. Its success hinged on this comprehensive collaboration. A close examination of Mancuso's donated materials reveals that while the architects stood at the project's most crucial points, the pavilion itself represents a collective achievement of both Korean and Italian citizens working together; a product of their combined intellectual and creative efforts.

In the early stages of discussions about the Korean Pavilion in 1993-1994, Korea and Italy's positions were not perfectly aligned. Korea sought to expand its cultural horizons globally through the Venice Biennale, while Venice focused on the public benefits the pavilion would bring. While the architect group represented by Kim Seok Chul and Franco Mancuso were the main players in

9. The opening ceremony date for the Korean Pavilion had already been set for June 7, 1995, and the completion inspection needed to be conducted by the end of May. However, the construction drawings for on-site delivery by the Mancuso and Serena studio were still being worked on until mid April.

reconciling these different interests and realizing what many believed impossible, it was the involvement of various departmental officials that helped shape the Korean Pavilion into a concrete manifestation of "sustainability," one of our era's highest values. The project's greatness lies in how it established and implemented design concepts of maximum respect for history and nature, and minimal environmental interference, even before systematic theories of global sustainability emerged in the architectural field during the early 1990s. For this reason, all who contributed to the pavilion's establishment in any way deserve recognition.

The year 2025 marks the 30th anniversary of the Korean Pavilion, serving as a time to reflect on its origins and celebrate the joint achievement of Korea and Italy. However, to call the Korean Pavilion at the Venice Biennale the brightest achievement in 140 years of Korean-Italian diplomatic relations may be premature, as challenges remain for both nations. These include developing systematic management and preservation strategies for the pavilion, and discussions on enhancing its utilization to serve as a regular venue for cultural exchange. Initiating these discussions is as crucial as the collaboration between the two architects who combined their wisdom 30 years ago to bring the pavilion to life. Just as their aspirations were realized, we believe our efforts will bear fruit. With hopes for an even brighter future built upon 30 years of memories of the Korean Pavilion, I conclude this writing.

WHEN WAS "THE LAST PAVILION"?

ALICE S. KIM

The Thirtieth Anniversary of the Korean Pavilion (1995-2025) and The Last Pavilion (ARKO: 2024)

Approaching the thirtieth anniversary of the Korean Pavilion (hereafter, KP) at the Venice Biennale (hereafter, VB), the pavilion's managing body produced a commemorative book titled *The Last Pavilion: The Archival Publication of the Korean Pavilion at the Venice Biennale*. As stated in the publisher's introductory remarks, "Arts Council Korea (ARKO) launches this archival publication, *The Last Pavilion*, in conjunction with the opening of the special exhibition *Every Island is a Mountain*," also organized for the anniversary at the 2024 Venice Art Biennale, "to reflect on its achievements and to redesign its future vision."[1] Whilst as "an archival publication," the stated aims of the book—of "reflection" on past "achievements" and "redesign" of "its future vision"—already indicate that the book's mandate far exceeds the nature of a mere archive.

Rather, the lengthy publication, numbering 325 pages in the Korean language edition and 443 pages in the English, is composed, on the one hand, of various kinds of archival material, including the text "Behind the Scenes: Designing the Korean Pavilion at the Venice Biennale" which chronicles in particular the "14 long months" of multiple design concepts and proposals and the back-and-forth with Venetian authorities by the late, Kim Seok Chul,[2] the principal architect of the KP, which was "discovered around 2012" in the ARKO archives but dates back thirty years in its writing.[3] But also included in the book are texts commissioned for the anniversary publication, such as the book's titular essay, "The Last Pavilion," written by the exhibition organizers, which reads like the official state-centric narrative of the KP,[4] alongside other retrospective essays by Franco Mancuso, the local architect and co-architect of the KP, translated from the Italian,[5] or select outside observers, namely art critics and curators, with intimate familiarity with the pavilion's activities, included in the section titled "Studies."

Needless to say, the mix of "archive" and "reflections" give way to an interesting array of corresponding and divergent narratives. While at first, they might

---

1. Choung Byoung Gug, "Foreword," in *The Last Pavilion: The Archival Publication of the Korean Pavilion at the Venice Biennale* (Arts Council Korea: 2024), 11.

2. For Korean names, I follow the Korean system of listing the surname first, followed by the given name. For the artist Nam June Paik, I use what has become the standard globalized usage.

3. Kim Seok Chul, "Behind the Scenes: Designing the Korean Pavilion at the Venice Biennale," in *The Last Pavilion*, 28-44. The unpublished text appears with a note indicating its discovery in the ARKO archives in 2012. Despite the clearly marked authorship by Kim Seok Chul and apparent date of writing around late 1994 to early 1995 based on its content, there is no publication record of the text from the period. Kim Seok Chul passed away in 2016.

4. Ho Kyoung-yun, "The Last Pavilion," in *The Last Pavilion*, 16-27.

5. Franco Mancuso & Ernesta Serena, "On the 30th Anniversary of the Venice Biennale's Korean Pavilion," in *The Last Pavilion*, 45-70.

appear to offer complementary fragments or differing viewpoints, on closer reading, crucial tensions emerge—in particular between the official state-centric narrative and the architect Kim Seok Chul's narrative—as to the meaning and significance of the KP. To be sure, the idea of the "last pavilion"—that is, the KP "as the last national pavilion in the Giardini"[6]—which forms the overarching theme of the book, is not presented with any sense of overt criticism or irony in either. Rather, it is couched within the generally shared storyline of overcoming the "impossible,"[7] in view of the competition from other nations in the face of strict building regulations and lack of available space in the Giardini. However, the official narrative of "The Last Pavilion" has the effect of both undermining and obscuring the substantial contributions of the artist Nam June Paik and the pavilion's principal architect Kim Seok Chul, and of more consequence, their ideals and visions for the KP, with significant implications for the contemporary mythologization of its past history as well as anticipation of its "future vision."

That is, as the South Korean state (hereafter, SK) is situated in the official narrative as the main protagonist of "The Last Pavilion," the spotlight is refocused from the "14 long months"[8] from Paik and Kim's initiation of the project to its ultimate approval by the Venetian authorities, to the "especially remarkable" speedy "mere seven months" from the "groundbreaking ceremony… to…completion" and its "successful opening,"[9] as the "small miracle"[10] of the KP is transplanted from the Venetian scene to the state-centric narrative as an echo of SK's economic-cum-cultural miracle motif. Not only that, but the artist Nam June Paik—whose formal appeals for the establishment of the KP, as well as his prodding and goading of Biennale officials and Venetian city authorities are well-documented—is relegated to the role of "midwife"[11] of the pavilion, while the role of the pavilion's architects Kim Seok Chul and Franco Mancuso, whose tenacious and sustained efforts to secure the pavilion, including no less than three different proposals, are nearly left out altogether, and Kim's narrative of events is both selectively appropriated and downplayed to the point of inconsistency.

These tensions, I show in this essay, are not only attributable to the divergent desires/aspirations and aims of their narrators, but also the patchwork of competing origin myths of the pavilion, namely that of the "last pavilion" and the "Korean" pavilion, which in the thirty-years hence have evolved into discrepant understandings of the significance of the KP. More specifically, there is a substantial and dangerous slippage in meaning that occurs from the concrete circumscribed idea of the "last pavilion" in Kim Seok Chul's proposal for the KP to the understanding of the "The Last Pavilion" in the official narrative in the anniversary reflections. As I examine the discrepancies that emerge between the two narratives, I focus my critique on the blind spots and contradictions that lie behind the official narrative's overlapping claims for international prestige and national primacy that ground its politics of legitimation, and how these strangely mirror and intersect the very conflicts and paradoxes behind the enhanced internationality of the VB since the 1990s. I argue that recounting and foregrounding the ethos and efforts of the architects Kim Seok Chul and Franco Mancuso, and artist Nam June Paik is important to keeping their visions for the

194

6. Choung, "Foreword," 11.

7. See for example, Achille Bonito Oliva, the director of visual art of the 1993 VB's initial response to the South Korean proposal to build a Korean pavilion: "It will probably be impossible." Kim, "Behind the Scenes," 31.

8. Kim, "Behind the Scenes," 41.

9. Ho, "The Last Pavilion," 22-3.

10. Kim, Behind the Scenes," 38; See also Jade Keunhye Lim, "Preface," in *The Last Pavilion*, 13.

11. Ho, "The Last Pavilion," 19.

12. Ibid., 19-22.

13. Oliva was appointed in 1992 but the opening of 1992 VB was delayed until 1993.

Korean Pavilion alive, and that those visions are crucial to its evolving mythology as well as anticipation of the potential realization of its future vision.

The "Last Pavilion": Deciphering its Origin Myths

The "Last Pavilion" and the "Korean" Pavilion
Inserted between the opening section on the historical significance of the VB and the national pavilions of the Giardini going back the late 19th-early 20th century, and the two closing sections which position the South Korean state and its postwar national economic development as the main protagonist of the KP in the official narrative of "The Last Pavilion," the section dedicated to recounting Paik's role, "The Midwife of the Korean Pavilion, Nam June Paik,"[12] presents in compressed form the background story of how the KP became a possibility, that is, changing from the "impossible" to possible. This is the shared theme that can be found across all the narratives contained in the book, whether archive or reflection, although with differing treatments. Arguably, it is the seat of the origin story of the Korean Pavilion that is the stuff of contemporary myth or legend.

The reconstructed storyline in this section follows the general sequence of events provided in more detail in Kim Seok Chul's text, up to the special luncheon hosted by the Korean Minister of Culture on the opening of the Daejeon Expo (August 1993) to which both Achille Bonito Oliva, who had been appointed Director of the Visual Arts section of the VB in 1992,[13] and Gino Di Maggio of the Fondazione Mudima in Milan, were invited.[14] From here, however, it abruptly veers off into a different account, alongside the introduction of a competing origin myth.

In Kim Seok Chul's account, this luncheon forms a crucial turning point for the KP in two ways: it was when "the project that had been discussed and propelled at the individual level had transformed into a government-level project,"[15] and second, because it was when the project got off the ground in relation to the Venetian side. At the luncheon, the South Korean side proposed to build the KP on the site of its current location between the Japanese and the German pavilions, designed on the "underground level so as not to disrupt the existing buildings and vegetation," which was received favorably by Oliva, going from "It will probably be impossible" to "let's give it a shot⋯I will do what I can,"[16] marking a key milestone in the KP initiative.

In lieu of this crucial turning point from which the architects' work began in earnest, the official narrative presents at this juncture the competing, counter claim that the idea for a shared North Korea/South Korea pavilion became the clincher that won over the Venetian authorities to propel the project forward.[17] This idea, of course, corresponds to the other main origin myth of the KP of the "Korean" pavilion, that is, Paik's idea for the joint North-South use of the pavilion that is best concretized in the sign that reads "Corea" (rather than Corea del Sud)[18] on top of the pavilion building.

From here, the official narrative telescopes nine months to "May 5, 1994, [when]

---

14. Ho, "The Last Pavilion," 20. See also, the more detailed discussion of the luncheon in Kim Seok Chul, "Behind the Scenes," 30-31.

15. Kim, "Behind the Scenes," 30.

16. Ibid., 31. The elements quoted here, i.e. the site location, underground idea, and Oliva's positive response, are also reiterated in Kim's fax to co-architect, Franco Mancuso, adding that Oliva said it was a unique and possible proposal. See fax from Kim to Mancuso on August 18, 1993, and also from Mancuso to Kim (in response) on September 10, 1993, Franco Mancuso Donated Archives, ARKO, Seoul.

17. Ho, "The Last Pavilion," 21. For this claim, Ho quotes from an external text written for the 2013 Korean Pavilion exhibition catalogue, by Yongwoo Lee, "Heroes of the Korean Pavilion at the Venice Biennale," in *Kimsooja, To the Breath: Bottari*, eds. Franck Gautherot and Seungduk Kim (Les Presse Du Reel, 2013), 13-18.

18. Ho, "The Last Pavilion," 21, footnote 4.

they submitted a formal application package for the construction of the pavilion," then picks up on German reunification, stating: "The slogan 'Healing political tensions through art' and the manifest desire for the eventual reunification of North and South Korea played a major role in winning approval for the Korean Pavilion," and ends with the Vice Minister of the Ministry of Culture and Sports "present[ing] the architectural conception of the pavilion and clarified the South Korean government's position on the Korean Pavilion as a symbolic project of a 'New Korea' aiming at globalization. As a result, the project was approved in a little over a year."[19] Condensed within the space of a page, it serves as a segue to the last two sections on South Korea's "brimming…international aspirations in the 1990s."[20]

Beyond the veracity of the competing claim, what is eclipsed in this official reconstructed narrative entails the bulk of Kim's text, the vivid exposition of "14 long months" of work and effort, including Kim's five trips to Venice, Mancuso's trip to Seoul, hundreds of faxes exchanged between them, Paik's letters and hand drawings for the more skeptical Venetian authorities, and so on. More to the point, it is out of the multiple proposals that the architects drew up throughout these negotiations with the Venetian city officials, fielding one obstacle after another, from whence the concept of the "last pavilion" concretely originates in its far more circumscribed yet substantial form—and where the origin story of the "Korean" pavilion also comes into play.

"The 'Last Pavilion' in the Giardini in its Centenary Year" (The Architects' Vision)

The "last pavilion" takes on a substantially different meaning in the pavilion's principal architect, Kim Seok Chul's text. As mentioned earlier, the initial proposal had been for an "underground pavilion," initially moving the project forward. However, further site surveys revealed "tree roots extending in all directions" making an "underground pavilion," which Kim and Mancuso planned to locate two meters underground, impossible.[21] Barring this option, they proposed a "transparent pavilion," which, through the use of columns, steel pile caps, capitals, and walls used from both sides, "the existing pavilions [in the Giardini] can be seen from the lagoon side," and indoor and outdoor spaces blended without distinction, preserving "the natural atmosphere and beauty of place."[22]

However, after another month's work on the second, transparent pavilion, "everything seemed to go back to square one with Venice having to elect a new mayor." Kim remarks, "By that point, I had almost given up."[23] Nam June Paik's support, however, remained steady:

> I received a call from Nam June Paik in New York almost every day. "Don't give up and go on. I will do whatever I can to help. Massimo Cacciari, who is likely to be elected, is with the Communist Party, which is concerning, but I have a plan. You have to keep working," he said.[24]

Here Paik's NK/SK joint-use "Korean" pavilion idea comes into play as Paik's

19. Ibid., 21.

20. Ibid., 22

21. Kim, "Behind the Scenes," 32. See also descriptions for the underground pavilion in fax correspondence between Kim and Mancuso, Franco Mancuso Donated Archives, ARKO, Seoul.

22. See also working proposals of the transparent pavilion shared between Kim and Mancuso, attached to the fax from Kim to Mancuso dated October 11, 1993, Franco Mancuso Donated Archives, ARKO, Seoul. Another element of an earlier version of the "transparent pavilion" proposal I found interesting was the idea of creating an enhanced rest area with public toilets and showers to complement the hundred-year-old building that they incorporated into the pavilion design that werewas originally used as public toilets.

23. Kim, "Behind the Scenes," 32.

24. Ibid., 33.

strategy to win over the newly elected Communist Party affiliated mayor Cacciari: "Paik sent a letter with a drawing to Cacciari. In the letter, he wrote something along the lines of 'It's an opportunity for you to be a Nobel Peace Prize laureate. The Giardini will celebrate its centenary next year, and if the only divided country in the world (South Korea and North Korea) with different ideologies participates to address the nuclear issue culturally, how significant and historical would it be?'"[25] Kim recounts that Paik also sent a letter and drawing to the skeptical "Director General of Urban Planning," who "happened to be a huge fan of Paik," and surmises that "the letter played a significant role in turning the tide."[26] Soon after getting past this hurdle, they came face to face with another, as the VB office proceeded to "put the brakes on our project" with the convincing point "that China may be prioritized in being granted the last pavilion in the Giardini."[27]

Then came "the final proposal." Kim writes: "For this proposal, I took a new look at the history of the Venice Biennale and the history of the city itself."[28] Taking into consideration the full weight of the history of the Giardini di Castello, which began as a park during Napoleon's occupation in the early 19th century, transforming into the Biennale site around the 20th, and was now only open for "three months every two years"[29] while "abandoned most of the time⋯entail[ing] a great challenge for the city authorities," they incorporated "a reform plan for the entire Giardini" into their final proposal for the KP, whereby, "construction and opening of the Korean Pavilion would mark a new beginning for the park itself."[30]

In the concrete, circumscribed meaning of the "last pavilion" as it emerged as part of this third, and final "visionary"[31] proposal for the KP within the design and approval process, I believe the part that needs most to be brought back to the contemporary consciousness and evolving mythologization of the KP (on its thirtieth year anniversary) is the part that completes the idea of the "last pavilion" as it was put forward in the proposal: that the "Korean Pavilion will be the 'last pavilion' in the Giardini in its centenary year, and it will also be the first pavilion that marks the beginning of the park's new century."[32] Much of the design concepts from the second, transparent pavilion, transfer over to this final proposal, which appears to be a deepening or maturing of its key concepts and its expansion horizontally to incorporate the entire Giardini to the city beyond it, while also engaging historically with the site. It is to mark the end of its last century, while spearheading the birth of the "park's new century." The ethos of the final KP proposal that the South Korean team submitted in April 1994 to the City of Venice for approval may be said to be that of using the KP as a bridgehead for the opening up of the Giardini to the city/world, or bringing the city/world into the Giardini more permanently, on a year-round basis, making the Giardini itself more open and transparent, both physically and symbolically. As Kim reiterated to the mayor, "It will not be an ordinary pavilion⋯but one that will⋯ awaken the Giardini."[33]

The architect's narrative also gives concrete context to the origin myth of the "Korean" pavilion. Though it may not have been the first mention of Paik's North-South joint use idea, it appears to have been far more consequential, combined

197

25-28. Ibid.

29. The architecture exhibitions were not regularized as a biannual event until the 2000s. See Klarissa Rosenfield, "A History of the Venice Architecture Biennale," *ArchDaily* (August 28, 2012), https://www.archdaily.com/267113/a-history-of-the-venice-architecture-biennale.

30. Kim, "Behind the Scenes," 33.

31. Ibid.

32. Ibid.

33. Ibid., 36.

with Kim's proposal for the "last pavilion" in this context. As Kim asserts, "Paik's letter with a hand-drawn image and our proposal to revive the Giardini seemed to have moved him," in reference to mayor Cacciari, who "agreed in principle" with the proposal idea, lending his support.[34] Kim's "last pavilion" and Paik's "Korean" pavilion thus appear equally as the imaginative visions arising from the common aim and process of getting past the "impossible." The visionary idealism-optimism of the artist Nam June Paik, that the "Korean" Pavilion might serve one day as a shared cultural platform for North and South Korea, is here matched and complemented by that of architect Kim Seok Chul, whose "last pavilion" was envisioned as the "first pavilion" to awaken both the park's natural greenery and dormant buildings long closed off to its surroundings, and as an integral part of a larger reform vision for the entire Giardini that also limns a symbolic departure from the imperialist and colonizing legacies of its first two centuries in which it was born,[35] through a design program for new spatial and aesthetic practices with which "to be born again"[36] into its next (post-colonial) century.

When Was "The Last Pavilion"? – Thinking at the Limit Between Postcolonialism and Global Capitalism

Insofar as it leaves out the concrete concept of the "last pavilion" that emerged out of Kim's third and final proposal for the KP, the broader symbolism of "The Last Pavilion" in the official narrative remains primarily held up by, and simultaneously upholding, the centrality of the Giardini (qua the seat of VB's tradition/national pavilions) as "the main stage of the Venice Biennale."[37] This centrality of the Giardini, however, is difficult to defend for its anachronism both politically and empirically, and at the same time, sits in an uneasy relationship to the nationalist trajectory of South Korean economic-cum-cultural development that forms the other main pillar of its narrative. In short, with the exception of the concrete, historicized sense of "the 'last pavilion' in the Giardini in its centenary year" within the circumscribed symbolism of the reform vision of the Giardini as presented in Kim's narrative, I argue that the Korean Pavilion has never been "the last pavilion" in any meaningful sense, even from its very outset in 1995, and that its upholding such a position is fraught with grievous contradictions. In this section, I take up the politics of legitimation behind the official narrative within the broader context of the evolving dynamics of the VB system, and the global expansion of contemporary art exhibitions since the postwar period especially emerging out of post-colonial countries of the Global South.

The Politics of Legitimation Behind the Official State-Centric Narrative of "The Last Pavilion"

The most straightforward expression of the "the last pavilion" as a distinguishing marker of the KP, grounded on the historical significance and international prestige of the VB, moreover, on its hundred-year anniversary, and on the centrality of the Giardini to the VB, can be found in the organizer's introductory remarks:

---

34. Ibid., 34.

35. This includes the older bicentennial history of the Giardini built during Napolean's occupation of Venice, ending centuries of Venetian independence, integrating it into the broader mold of 19th century European imperialism and Age of Empire. On the history of the Venice Biennale inheriting the mold of the imperialist world's fairs that came before it, see Caroline A. Jones, *The Global Work of Art: World's Fairs, Biennials, and the Aesthetics of Experience* (University of Chicago Press, 2016), Chapter 3 (Old World/Biennial Culture).

36. Kim, Behind the Scenes," 33. Kim introduces his reform plan thus: "This historic site of international contemporary art that will celebrate its 100th anniversary is walking the path of its fall. It needs to be born again on the occasion of its centenary…"

37. Lim, "Preface," 13.

Founded in 1895, the Venice Biennale stands as the world's oldest and most prestigious international art exhibition, comprising the main exhibition led by curators and national pavilion exhibitions representing individual participating countries. Since opening the Korean Pavilion as the last national pavilion in the Giardini, the main venue of the Venice Biennale centennial in 1995, Arts Council Korea (ARKO) has served as a bridgehead to introduce South Korean art to the global stage for the past 30 years.[38]

"The Last Pavilion,"[39] the publication's titular essay, further advances this narrative on two fronts. The first is the line that is drawn from the establishment of the VB in the Giardini in 1895 and the "Belgian Pavilion⋯the first to open under this system in 1907⋯followed by the Hungarian Pavilion (1909), the German Pavilion (1909)," and so on, to "the Korean Pavilion," which "South Korea constructed⋯in 1995 and remains the last country to open a pavilion in the Giardini."[40] The second is a separate, parallel line drawn from postwar "South Korea's economic development" and the rise of "international events hosted by South Korea, including the 1986 Asian Games, the 1988 Seoul Olympics, and the 1993 Daejeon Expo" to "the rapid increase in overseas activities and international exchanges taking place in South Korean art in the 1990s."[41] Within this trajectory, "the year 1995," designated the "Year of Fine Arts" as part of "South Korea's ten-year cultural development plan," saw not only the opening of the Korean Pavilion at the VB but also South Korea's first art biennale, "the Gwangju Biennale," inaugurating "a new system of biennales⋯in the South Korean art world."[42]

In advancing the aspirations for both international prestige and national recognition, ARKO's official narrative of "The Last Pavilion" rests on the split between (and superimposition of) two contrasting lineages—one (at the supranational level) that moves from the 1995 Korean Pavilion to late 19th century Europe, and the other (at the national level) to postwar South Korean industrialization and the economic miracle of the 1970. That is, its overlapping claim for internationalism and nationalism is fraught with much of the same "fragmentation" and "asymmetry" characteristic of the "international nationalism" of the Venice Biennale, about which Italian historian Clarissa Ricci cogently argues.[43] It operates on the same logic or "politics of legitimation" which in the case of the Venice Biennale, uses "the continuous rise in the number of pavilions" as the "'international' parameter to measure the success of the exhibition,"[44] while in the case of the official narrative of "The Last Pavilion," uses not only VB's increasing pavilions but also its long, centennial history (i.e. the cultural capital that accrues in old spaces of world culture[45]), alongside South Korea's growing internationalism as the "parameter" to measure the breadth and caliber of its national success or greatness.

Working from Ricci's critique, "internationality" in the VB (as in the official narrative of the KP) operates mainly with recourse to national representation, as seen for example in the VB president Paolo Baratta's maintenance of the age-old national principle in 2011 in claiming the rise in pavilions as "clear evidence

199

38. Choung, "Foreword," 11.

39. Ho, "The Last Pavilion," 17-27.

40. Ibid., 18. In this listing of pavilions, the fact that the earliest pavilions were also some of the largest European empires, and that aside from the US, former colonial countries were only added after WWII, alongside Japan (i.e. not a former colony but also not quite a European power), should indicate that this is not a simple chronology reflecting a universalist internationalism.

41. Ibid., 23.

42. Ibid., 25.

43. Clarissa Ricci, "From Obsolete to Contemporary: National Pavilions and the Venice Biennale," *Journal of Curatorial Studies* 9, no.1 (2020), 28-29.

44. Ibid.

45. For a discussion on the concept of the economy of prestigee in the field of world literature, see Pascale Casanova, *The World Republic of Letters* (Harvard University Press, 1999).

of growth and a 'pillar' of the exhibition," in the same way the Venetian mayor "Filippo Grimani did in 1895."[46] Hence, while artistic directors or curators of the VB have honed in on the critical and often self-reflexive approach to internationalism or globalism (e.g. "nomadism" which was highlighted in the 1993 VB curated by Oliva) since the 1990s and into the 2000s within the Biennale's main thematic exhibitions (e.g. Oliva, Jean Clair, Harold Szeemann, Okwui Enwezor, Massimo Gioni, etc.), and including even the case of some national pavilions (such as Klaus Bussman who invited Hans Haacke and Nam June Paik to represent the German Pavilion in 1993)—the Biennale board, Ricci argues, continues to maintain its anachronistic and conflicting position espousing both "international reach and national primacy" while "shift[ing] the discourse to the complexities of⋯globalization."[47]

The problem with such a politics of legitimation for the VB is the paradox of proliferating national pavilions, especially since the mid-1990s, in the face of both internal and external criticism of the principle of national representation and calls for abolishment of the national pavilions structure since the late 1960s.[48] Such criticism, however, were never dealt with but merely managed with VB's push to enhance its "internationality" with its turn to "the transnational approach" in the 1990s, thus effectively halting calls for the pavilions' abolishment.[49] As a means of placing the spotlight on "the main international exhibition, curated by the Biennale" for the Central Pavilion (i.e. the former Italian Pavilion), this approach "gave a more appealing⋯framework to the outmoded national pavilion system" thus finding "a way to incorporate its traditional system in a contemporary setting and in return, continu[ing] to receive formidable economic support from the pavilion-owning countries."[50]

On the other hand, in the case of "The Last Pavilion," the problem can be seen in the intersecting paradox of ① the anachronism of holding up the centrality of the Giardini (the seat of tradition/national pavilions) as part of its own politics of legitimation, in the face of sustained political critique of VB's national pavilion system and contravening empirical events; and relatedly, ② the dissonance or contradiction between this position and its parallel national economic-cum-cultural development trajectory, in which the Gwangju Biennale, also founded in 1995, is generally perceived to be an active part of the new group of post-war biennials emerging from post-colonial countries of the Global South, challenging the cultural hegemony of old Eurocentric structures and practices inherited from imperial and colonizing legacies such as the 19th century world fairs and international exhibitions. Both positions pose critical implications for the politics of the present position and future vision of the Korean Pavilion.

"Moving Out of the Giardini" and the 1995 Venice Biennale
As much as the national pavilions have served as the "trademark feature" of the Venice Biennale, they have also remained its most "controversial feature," situated at the heart of the paradox surrounding its conflicting internationalism.[51] As Ricci recounts, spurred by postwar internal reforms for democratization within the VB (seeking to expunge the imprint of its prewar policies under Fascism)

46. Ricci, "From Obsolete to Contemporary," 28.
47. Ibid., 25-9.
48. Ibid., 18-21.
49. Ibid., 21-9.
50. Ibid., 29.
51. Ibid., 9, 29.

in conjunction with external criticism of the national pavilion system coming out of new postwar biennials in Europe as well as the Global South, the push to "mov[e] out of the Giardini" began in the late 60s and 70s.[52] Following the 1968 student movements, not only was "the issue of the obsolescence of the national pavilions" discussed within internal debates seeking a reformed charter for the Biennale, "known as 'the reform of 1973'," "as a new generation of Italian curators, including Germano Celant and Gillo Dorfles, advocated their abolition," but even some reform-minded commissioners of the national pavilions also "believed that the national pavilions were 'fossils'."[53] This viewpoint continued into the 1980s in Italian discourse with "national representation⋯perceived as part of the legacy of the nationalistic chauvinism of right-wing politics" on the backdrop of the formation of the EU and rise of globalization.[54]

As for its external critique, the same discussions emerged in the 1970s at the Sao Paolo Biennale (1951), the earliest post-colonial biennial outside of the West. While the Sao Paolo Biennale did not have "constructed pavilion buildings, the organizers⋯had difficulty in avoiding pavilion-based national representation because of the economic contribution of each country."[55] This followed closely the VB model of national representation which developed from "each member nation both funding and selecting their artists," later turning into "independent pavilions managed autonomously by countries" also for economic reasons.[56] Nonetheless, the Sao Paolo Biennale began to move away from this structure in 1979 with its first Latin American Biennial, to the eventual abolishment of national pavilions in 2006.[57] This move was further echoed in new Southern biennials coming into being explicitly rejecting the pavilion model, such as the Sydney Biennial founded in 1973.[58] On the other hand, the Havana Biennial (1984) with its explicit orientation toward the Global South, by its third, historic, edition, not only did away with national representation and the awards structure, fully eliminating "the last remnants of the Venice Biennial model," but also went on to invite "diasporic artists living in the global centre" alongside artists from the global periphery, expanding the concept of the Global South, according to Mosquera, its head curator, while broadly challenging the European system.[59]

The VB reforms for wide-ranging democratization within which the issue of the national pavilion system was central were slow going, and thus "moving out of the Giardini" only became "normal practice" in the 90s, due to the "inevitable consequence of the fact that more countries had started renting buildings as temporary national pavilions in the city of Venice" which developed as "a practical solution" to the pressure to host the growing number of participating countries in the Central Pavilion.[60] In fact, the Taiwan Pavilion, which was also established in 1995, the same year as the KP, was one of the first national pavilions to be established outside of the Giardini.[61] Since 1995, The Taipei Fine Arts Museum, the managing body of the Taiwan Pavilion, has rented the Palazzo delle Prigioni, a sprawling 200 square-meter former prison building next to the centrally located Basilica di San Marco.[62] Moreover, it was the 1995 VB in which Jean Clair, the first non-Italian invited outside curator, placed "the main exhibition outside the Giardini⋯The 46th [Biennale], entitled *Identity and Alterity*, was in fact, held

---

52. Ibid., 15-19, 26.

53. Ibid., 18-19.

54. Ibid., 20.

55. Ibid., 19.

56. Ibid., 11, 13.

57. Ibid., 19.

58. Ibid.

59. Oliver Marchart, "The Globalization of Art and the "Biennials of Resistance: A History of the Biennials from the Periphery," *On Curating* 46 (June 2020); https://www.on-curating.org/issue-46-reader/the-globalization-of-art-and-the-biennials-of-resistance-a-history-of-the-biennials-from-the-periphery.html.

60. Ricci, "From Obsolete to Contemporary," 25-6.

61. Stephen Naylor, *The Venice Biennale and the Asia-Pacific in the Global Art World* (Routledge, 2020), 14, 118.

mainly at the Palazzo Grassi," explains Ricci, to the dismay of Biennale board members.[63] Then, after the VB finally got its new charter in 1998, completing the reforms from 1973, its 1999 relaunch marked the most significant move "out of the Giardini" into the Arsenale:

> The 48th Biennale d'APER-Tutto (1999), which has been greatly celebrated for its expansion into the Arsenale, was not transnational, but global. Szeemann firmly resisted the idea of national pavilions; however, he ultimately only had authority over the Italian pavilion. He therefore moved Italian artists from the Central Pavilion and dispersed them throughout the main exhibition.[64]

As these events surrounding the opening of the Korean Pavilion in 1995 show, just as the KP was being added as the "last pavilion" of the Giardini, the VB was moving outside its boundaries, both due to the growing number of participating countries and the critical approach of VB curators, including their pointed pushback on VB's pavilion system. It is hard to know how much Kim Seok Chul was aware of VB's long controversy over its national pavilions, but his proposal for the "last pavilion" can be seen to be in line with this tendency towards opening up and moving out of the restrictive boundaries of the Giardini. The Korean Pavilion was not unique, insofar as it was part of the larger growth of national pavilions in the 1990s, amidst charged conflicts between VB curators and board members over the terms of VB's "internationality," wherein the centrality of the Giardini as the seat of the VB became a key point of contention. Although arriving at it from different angles, the Korean Pavilion, like the Taiwanese Pavilion, is embedded within the same paradox.[65] As such, national pavilions have continued to be established every year since (1995), and increasingly so, within the Arsenale as well as spread throughout the city of Venice. For example, as of 2015, out of 89 total national pavilions, 29 were located in the Giardini, 29 in the Arsenale, and 31 across the city.[66] The recent news that Qatar may be planning on building a national pavilion in the Giardini is less consequential to challenging the idea of the KP as "The Last Pavilion" presented in the official narrative than this more substantial history of VB's growing pavilions vis-a-vis controversies and ongoing debates over its traditional national pavilion structure. As with the Taiwanese Pavilion, which presents quite an interesting difference/deferral to the KP, the "last pavilion" has been and continues to be deferred by the ever-growing number of new national pavilions since it first came into being in 1995.

At the same time, upholding of the centrality of the Giardini is contradictory on the broader plane of SK postcolonial politics, and runs up against South Korea's own Gwangju Biennale, widely perceived as constituting part of the wave of new postwar contemporary art exhibitions, from Sao Paolo to the Havana Biennial, the Asia-Pacific Triennial, Johannesburg, etc., shaping "a biennial of resistance."[67] The fetishization of the center, in the face of South Korea's participation, via Gwangju, in the discourse of what Ranjit Hoskote has famously called "the emergence of a global South"[68]—as part of a marked group of post-

---

62. Chu-Chiun Wei, "From National Art to Critical Globalism: The Politics and Curatorial Strategies of the Taiwan Pavilion at the Venice Biennale," *Third Text* 27, no. 4 (2013), 475.

63. Ricci, "From Obsolete to Contemporary," 25.

64. Ibid., 26.

65. According to Stephen Naylor, "Rumour has it that the potential site behind the Japanese Pavilion had been raised as a possible site for a New Taiwan Pavilion, although this site was quickly secured by Korea."

The Venice Biennale and the Asia-Pacific in the Global Art World, 117.

66. See the VB website: https://www.labiennale.org/en/art/2015/biennale-arte-2015-all-worlds-futures; Art Council Malta website: https://artscouncilmalta.gov.mt/archive/the-world-in-one-city/

67. Ranjit Hoskote, "Biennials of Resistance: Reflections on the Seventh Gwangju Biennial," in *The Biennial Reader*, eds. Elena Filipovic, Marieke van Hal and Solveig Ovstebo (Hatje Cantz, 2010), 310.

colonial countries of the Global South having broken through the gates/spaces of the rarified cultural capital of globalized contemporary art exhibitions, forming the parallel nationalist genealogy of the KP alongside Gwangju via postwar SK industrialization and development—presents a glaring contradiction. That is, the deeper trouble with the official narrative of "The Last Pavilion" is that it exposes the reconstructed identity (and thus also the future vision) of the KP to political anachronisms and contradictions owing to key blind spots that lie behind its overlapping, conflicting claims for international prestige and national primacy that ground its politics of legitimation.

A Requiem for Nam June Paik's and Kim Seok Chul's Visions for the Korean Pavilion

What becomes especially apparent in the respective yet intersecting contradictions of the position of the VB board members and the official narrative of "The Last Pavilion" is the shared context of new postwar developments in the geoeconomics and geopolitics of global capitalism, and the differential locations of the VB (as the seat of old power/capital) and the KP (as part of the new East Asian locus of global financial power since the late 70s and 80s into the "globalization" decade of the 90s). In particular, I would point to the overlapping crises of: ① "the crisis of the uncompleted struggle for 'decolonization'"[69] for much of the once-colonized and once-colonizing world, in the postwar period, accompanying the transformation from Roosevelt's politically revolutionary "one-world" vision for post-WWII world governance to the postwar US-led Cold War order, wherein the promise of the equality of nations as represented by the emerging UN became primarily instrumental to US global hegemony[70]—of which the divided Korean peninsula remains a particularly potent symbol; and ② "the crisis of post-independence states," namely the "persistence of many of the effects of colonization, but at the same time their displacement from the colonizer/colonized axis to their internalization within the decolonized society itself."[71] Here, for example, the postcolonial state holding up outmoded systems/structures of economic and cultural power of old imperial centers, such as the VB pavilion structure with its own politics of legitimation.

The broader political background for these intersecting contradictions— of VB's "international-nationalism" evidenced in growing national pavilions, which now increasingly include those once-colonized from late, industrialized or industrializing global peripheries or semi-peripheries (such as the economic powerhouses of SK and Taiwan in the case of the KP and TP) building or renting national pavilions since the 1990s, on the one hand; and the "international-nationalism" of KP's official discourse of "The Last Pavilion" on the other, which upholds/rests on the centrality of the Giardini and the VB as the "oldest and most prestigious international art exhibition"[72] as part of the politics of its own legitimation, both in the face of longtime internal and external, global critique of VB's national pavilion system seen to "perpetuate a retrograde political and power

---

68. Ibid.

69. Stuart Hall, "When was 'The Post-colonial'? Thinking at the Limit," in *The Postcolonial Question: Common Skies, Divided Horizons*, eds. Iain Chambers and Lidia Curti (Routledge, 1995), 244.

70. Giovanni Arrighi, *The Long Twentieth Century: Money, Power and the Origin of our Times* (Verso, 2010), 67-70.

71. Hall, "When was 'the Post-colonial'? Thinking at the Limit," 248.

72. Choung, "Foreword," 11.

system"[73] and calls for its abolishment coming out of the 1968 global student protests or as part of the counter-hegemonic critique of Western biennale models accompanying the rise of new biennials outside the West, such as the "biennales of resistance"—can be traced in large part to these two ongoing crises conditions. Moreover, it is also these ongoing crises of postcolonialism at the intersection of global capitalism that not only form the ground or condition of possibility, but also the staying power for the mythologization of the origin stories of the "last pavilion" and the "Korean" pavilion.

In January 1995, ahead of the opening of the Korean Pavilion at the Venice Biennale, Nam June Paik is said to have said the following:

> It was a shame not to have a Korean Pavilion, but there is no need to be arrogant about having one. We have now gone from a backward country to an average one, but if we think that we have become a culturally advanced country because of it, we will become nothing but a laughing stock. We should also not think that that is the only way to become first rate. If we make a fuss like it is the Olympics, it will be an embarrassment internationally. The Olympics and art are different. In sports, it is important to be first, but in art, the key issue is not who is better, but how we are different.[74]

Paik's words ring true today on the thirtieth anniversary of the KP as when he first uttered them in 1995. As the art critic Young-chul Lee ends his reflections on the KP with a sustained focus on what might be called Paik's critical globalism with these words, he aptly introduces the quote with a warning to "ordinary people, artists, art councils, and officials of the Ministry of Culture, Sports and Tourism alike," to heed Paik's words and safeguard the KP from becoming "a place of rigidity, prestige, and pretensions."[75] My engagement with the pavilion's origin stories as well as their evolving mythologies is also in line with this sentiment. I would add that Paik's pithy reproach also reads like an aphorism that gets to the heart of the issue of the postcolonial predicament of national pavilions (as a vehicle of national representation) within the context of globalized contemporary art exhibitions, developing out of the conditions and constraints posed by the geoeconomics and geopolitics of global capitalism.

Despite the labored petitioning and approval process for the KP, the proposal to open up the Giardini on a permanent basis—and for the Korean Pavilion as the "last pavilion" of its centenary year to serve as the "first pavilion" to spearhead the Giardini's "new century"—still has yet to materialize amidst the context of growing national pavilions alongside the expansion of contemporary art biennial culture. At the same time, the idea of the joint use North-South "Korean" pavilion and the desire for reunification of the divided Peninsula behind it remains for many living within and beyond the Korean peninsula today as impassioned as it had been back then. However, as we find ourselves in the face of the near full reversal of any gains from the Sunshine Policy of the early 2000,[76] and more recently, caught in the firing lines of a new Cold War with China (rather than the

---

73. Ricci, "From Obsolete to Contemporary," 29.

74. Nam June Paik quoted in Young-chul Lee, "How the Korean Pavilion Came To Be," in *The Last Pavilion*, 374. I have re-translated the quote from the original Korean text from the Korean edition (270), to better convey the meaning of the original quote.

75. Ibid.

76. SK's Sunshine Policy itself came out of the very gains of South Korean democratization in the late 1980s after decades of US-backed authoritarian rule that characterized the bulk of the period of South Korean development.

former USSR as the primary target) waged by the US since the 2010s, bracing to avoid a historical repetition of hot wars across the region, Paik's vision of the "Korean" Pavilion appears no closer today to becoming actualized than it was during the volatile 1994 nuclear crisis between North Korea and the US when it was first proposed by Paik.

These are now the tasks of current and future generations to take up, and to these efforts going forward, I would recommend they continue to revisit Kim Seok Chul's concrete, experiential narrative of the "last pavilion" in considering the past as well as the future of the Korean Pavilion, because in addition to its healthy dose of visionary idealism-optimism, it provides a far more substantive form of engaging with and being part of that history.

SUSTAINABILITY, ARCHITECTURE AND THE VENICE BIENNALE

SONG RYUL, CHRISTIAN SCHWEITZER

"We want the tools and legal guarantees to carry out qualified work that can insert us critically into society."
— Protest sign during the opening days of the Venice Biennale in 1968

The Venice Architecture Biennale has always served as a mirror of the zeitgeist, bringing together local topics of 63 national pavilions and a central exhibition exploring a shared theme to an international stage. It stands as the largest and most influential gathering of architects worldwide, offering a unique platform for exchanging ideas and shaping the global architectural discourse.

In recent years, this discourse has been increasingly dominated by sustainability, highlighting its urgency and importance. For instance, in 2021, the United Arab Emirates Pavilion won the Golden Lion for its research into a renewable building material offering a sustainable alternative to Portland cement. Similarly, in 2023, the Brazilian Pavilion received the Golden Lion for its exploration of traditional rammed earth construction techniques. Beyond technical innovations, many pavilions have focused on social sustainability. Notable examples include the Philippines Pavilion in 2021, which received a special mention for its hands-on project on community self-organization in response to the climate crisis, and the Korean Pavilion that same year, which provided an open platform for interdisciplinary discourse involving over 200 contributors.

Discussions about the sustainability of the Biennale itself have also grown louder. Proposals for zero-waste exhibition concepts, the use of locally sourced materials and labor, engagement with local communities, and repurposing exhibited installations reflect an increasing awareness of the environmental impact of the event. The Korean Pavilion in 2021 was designed under the premise that every single material or item either had a use value for repurpose or was compostable, to the effect that nothing had to be junked or shipped back to Korea and got distributed to local artist communities after the closure of the event.

By nature, however, a temporary global exhibition like the Biennale can never be fully carbon-neutral, aside from relying on insufficient offset schemes.

Solely the carbon footprint generated by the travel of participants, visitors, and the transport of installations to Venice already is substantial. In theory, only a radical degrowth of the Biennale would achieve true sustainability, but this would contradict its fundamental purpose of showcasing the work of the global architecture profession. Maybe it is necessary to reconsider and redefine this purpose: how can the Venice Architecture Biennale contribute more directly and effectively to addressing the critical issues of our time?

*

Architecture is responsible for 38% of global greenhouse gas emissions and for 60% of the global waste production. As the single largest contributor to the problem, architecture is in urgent need for drastic emission reductions. Cities like London, Madrid, and Seattle are projected to experience a temperature rise of 6 degrees, New York 4 degrees, and Seoul and Tokyo 3 degrees Celsius. None of these cities are prepared for these changes, and a significant portion of their existing building stock will soon become uninhabitable.

This presents an additional challenge to the built environment: now we must prevent climate change itself and at the same time fight the impact of climate change. Unfortunately in recent years the discourse in architecture on the sustainable city got mixed up with the discourse on the resilient city. We tend to prioritize the problems we can see and feel in person over the similarly urgent problems that are not tangible. For most architects today, when they speak about the sustainable city, they are actually referring to the resilient city. Googling "sustainable architecture" for example will predominantly show results addressing the resilient city with only few results on the sustainable city. A solution to a heating city might sometimes peripherally contribute to a reduction of carbon emissions from building material or energy consumption, but in most cases it requires efforts that stand in direct contradiction to the requirements to achieve sustainability. While both are equally important, we have to distinguish both clearly to not lose focus on one of them.

To decarbonize the atmosphere, combat climate change, and cool cities, intact ecosystems are essential. Forests and peatlands serve as vital carbon sinks, absorbing more carbon than they emit. However, global forest loss is still on the rise with nearly 50% above the level needed to meet the UN Zero Deforestation Pledge, making this goal de facto unattainable. With rising temperatures across the EU, land carbon sinks have diminished, with annual absorption dropping by a third in the last decade. The land sink of the Nordic Region has already vanished and its forest sink shrank by 90%, turning its land sector into a net carbon emitter, actively contributing to climate change, and thereby jeopardizing the continent's climate goals.[1] This issue has global implications, as at least 118 countries, including Korea, rely on natural carbon sinks to meet the UN Paris Agreement climate targets. We are simultaneously exploiting and destroying nature while depending on it to save us.

1. Patrick Greenfield, "What Happens to the World if Forests Stop Absorbing Carbon? Ask Finland," *The Guardian* (October 15, 2024).

Sustainability, in its broadest sense, refers to the ability to maintain or support a process continuously or indefinitely. We must ask ourselves: What process are we referring to? Is it the planet's ability to sustain life, or is it the continuation of human life itself? The current push to protect nature to solve our problems is beginning to prioritize nature over human life. Nature protection programs for carbon offset schemes, which involve fencing off parts of the rainforest, have already led to indigenous people being cut off from their native land, depriving them of their ability to sustain themselves. Nature protection is becoming another tool of neo-colonial capitalism, advancing the interests of the developed world at the expense of vulnerable parts of global society.

To change our relationship with nature, we must first transform our inter-human relationships, acknowledging that every human life has an equal right to exist and ensuring its protection. This requires ending wars, abandoning selfish national interests, ensuring equal access to clean water, food, shelter, resources, and energy, and achieving full equality among all human beings. Only then can we extend these principles to nature, and make them truly effective.

*

Looking at the history of architecture through a different lens, we can view it as a history of sustainability, especially since the pollution of our environment became evident during the industrial revolution in the 19th century. Buckminster Fuller phrased it as "progress through fear."

Modernism began with the goal of creating "healthy, social, and economic" architecture, as outlined in the 1928 CIAM La Sarraz Declaration. With the rise of the environmental movement in the 1960s, fueled by the energy crisis and the anti-nuclear movement of the 1970s, architects began promoting "ecological, social, and economic" architecture. In 1987, the United Nations defined sustainability in its Brundtland Report, outlining the Sustainability Triple Bottom Line, as the balance of the "environmental, social and economic" impact of any projects so we can meet the needs of the present without compromising the ability of future generations to meet their own needs. This definition expands the scope of architecture beyond 20th-century developments, requiring a balance between current and future needs. By the early 1990s, we had a comprehensive understanding of ecological and sustainable architecture at a theoretical level and at a technical building level.

In a simplification of this definition we understand sustainable architecture today as architecture that seeks to minimize the negative environmental impact of buildings through improved efficiency and moderation in the use of materials, energy, development space, and the ecosystem at large.[2] There are six basic strategies for sustainable building design, following the "R's" of sustainability, ranked by their effectiveness in reducing carbon emissions from low to high—recyclable materials, renewable materials, reusable materials, reduction of energy consumption, refurbishment of existing buildings, and refusal to build. Refusal,

2. Wikipedia, Sustainable Architecture, 2005-2025.

as the most effective strategy, although counterintuitive to many architects, aims to avoid constructing new buildings altogether through a strategic approach of reorganizing existing space. This strategy requires intelligent reinterpretation of function and program, a stringent adaptive reuse of the existing building stock and a strict moratorium on the demolition of existing buildings. As Anne Lacaton famously said: "Never demolish."

What all these strategies share is the concept of time. We must extend the lifespan of materials and buildings while minimizing energy consumption. The goal is to design four-dimensional building systems and strategies where the lifetime is core to their concept. Unfortunately, no building constructed in recent years in Korea fully aligns with any of these strategies. This is not due to a lack of awareness among architects but rather the result of systemic issues, including current mechanisms in the building industry, the absence of sufficient sustainable building laws, and a lack of political and legal frameworks that prioritize sustainability over growth.

*

By coincidence of history, the Korean Pavilion in the Giardini happens to be the most sustainable of the national pavilions on a technical building level. Built in 1995 as the last addition, it had to be designed under strict guidelines to minimize its impact on the park. Architects Kim Seok Chul and Franco Mancuso initially proposed an underground pavilion, but it was recognized that being out of sight does not necessarily equate to minimal environmental impact. Instead, the design had to preserve the vegetation and topography, reflecting a growing awareness since the 1980s of the need to protect the natural landscape in European cities. By the 1990s, removing a healthy tree had become nearly impossible to approve.

The resulting structure is a lightweight steel design resting on low-impact point foundations, hovering above the ground and weaving between existing trees. This approach allows the pavilion to be easily dismantled, reassembled at another location, or its materials separated and almost entirely recycled. The demand for minimal environmental impact naturally led to the application of sustainable strategies, demonstrating how an appropriate legal framework can drive sustainability—even when it is not the architects' explicit intention, as evidenced by their initial underground proposal.

The Korean Pavilion was only granted final approval after Korea assured the mayor of Venice that it could serve as a social space, such as a café, in addition to its function as an exhibition venue. This led to the inclusion of a water connection and a toilet, making the pavilion highly adaptable to changing uses and programs. This flexibility adds a dimension of social sustainability that is rare among the pavilions in the Giardini, making the pavilion a truly four-dimensional building, where time is inseparably incorporated into its concept.

*

Social sustainability involves designing physical spaces and social systems that meet people's primary needs. It is defined by equity—the fair distribution of resources, both physical (e.g., food, water, shelter) and social (e.g., access to information and freedom of speech); well-being—the provision of essential services like healthcare, education, and housing; and diversity, social justice, and democracy, ensuring everyone has a voice and a role in decision-making.

New Zealand and Finland were the first countries to implement a revised budget policy requiring all new spending to advance one of five government priorities focused on improving citizen welfare. These priorities include enhancing mental health, reducing child poverty and inequalities, thriving in a digital age, and transitioning to a low-emission, sustainable economy. This approach represents a shift away from traditional economic measures like productivity and growth, instead prioritizing community, cultural connection, and equitable well-being. It can be understood as the most radical example of social sustainability.

South Korea is ranked 14th globally by GDP out of 195 nations, yet, its social performance tells a different story. In the World Happiness Report, Korea ranks only 57th out of 150 nations. Similarly, in the OECD Better Life Index, Korea ranks 34th out of 41 nations. The country performs last in gender equality, birth rate and suicide rate, and at the bottom section in life satisfaction, social equality, and income equality. Despite its economic success, widespread dissatisfaction persists, reflected in the popular term "Hell Joseon." These findings underscore the inadequacy of GDP as a sole measure of national success, failing to capture quality of life, societal progress, and overall well-being.

Compounding these issues, South Korea remains the only developed country with still rising carbon emissions and ranks among the top three carbon emitters per capita. In contrast, Finland—despite having a similar GDP per capita—has been named the world's happiest country several years in a row by the World Happiness Report. Finland also leads globally in carbon emissions reduction, achieving a 43% decrease in 2022, and thereby the aim of the Paris Agreement 8 years ahead of time, while most nations have already postponed their compliance to 2050. This suggests a strong correlation between societal contentment and the willingness to embrace and implement change in the face of a looming geocide.

These observations highlight the necessity for architecture to broaden its focus from technical sustainability to a more comprehensive, holistic approach that incorporates social sustainability—a shift from architecture as merely the creation of new structures to architecture as a social strategy focused on reorganizing and improving existing urban environments. In order to make a meaningful impact, architecture must directly engage with the reorganization of society, bypassing the traditional role of buildings. Throughout history, many architects have looked beyond the narrow scope of architecture, focusing on broader societal strategies. Now, in a moment when it is clear that we must radically redefine our way of life, along with our social, economic, and political systems, architecture—being part of those systems—must also redefine itself. Architecture must be seen as the critical commentary on the human-made environment. It should be a tool for organizing society, where the creation of buildings is a secondary concern; in other words: "strategic architecture."

*

However, efforts of social sustainability within the Korean Pavilion seem futile—and were ultimately never implemented—given that the Giardini, one of the few park areas in Venice, is not freely accessible to the public. During the Biennale, access to the park is tightly controlled and limited to paying visitors, closing every day between seven and eight pm, while in the winter months, the park is locked down entirely. In 2023, the Austrian Pavilion sought to address these issues by engaging with the adjacent residential neighborhood to make the Biennale more inclusive. Their proposal split the pavilion into two parts: a public space freely, and for free, accessible via a pedestrian bridge over the enclosure wall, featuring a stage for events, and an exhibition area accessible only through the Biennale grounds. However, the Biennale Foundation rejected the plan, and only the bridgehead on the Giardini side as a symbolic protest against the project's refusal was executed.

Perhaps it is time to return the Giardini—effectively privatized by the Biennale—to the citizens of Venice as a public commons. In the face of rising temperatures, it could serve as a vital relief area with a broader cultural and recreational program. This shift aligns with the need to rethink the outdated concept of national pavilions. At a time when global cooperation and equality are essential for survival, the presence of 30 privileged nations with individual pavilions in the Giardini feels anachronistic. Similarly, the restrictive rules upheld by some pavilions, such as the requirement that only U.S. citizens can represent the United States Pavilion, seem counterproductive. A more inclusive approach could involve relocating the Biennale entirely to the Arsenale grounds, with national exhibition areas distributed annually through random selection.

The first Golden Lion for a national pavilion was awarded only in 1986, as part of the reinstated Gran Premi, which had been suspended in 1968 following widespread protests. Before that, the Gran Premi was awarded exclusively to individual participants. Instead of awarding a single national pavilion, recognition should go to collaborative projects or shared themes across multiple pavilions, fostering the transnational discourse and cooperation needed to address global challenges.

*

As an individual architect, the only way I can resist the unsustainable logic of the global building industry is by refusing to build. Yet even this act would

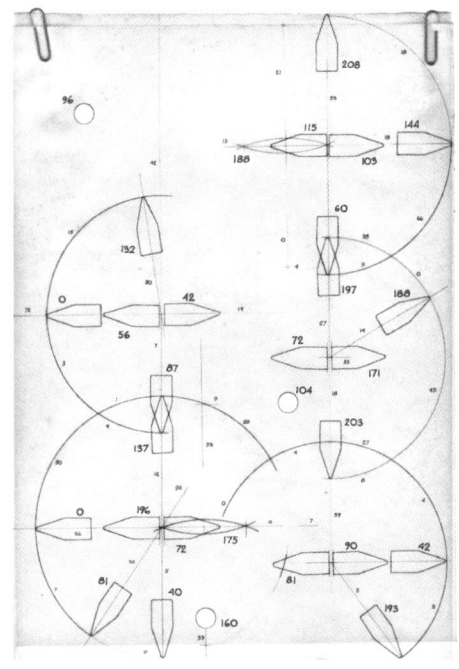

↖ Song Ryul, Christian Schweitzer, Anatomy of the Korean Pavilion, 2024
↑ Song Ryul, Christian Schweitzer, The Principles of Social Sustainability within the Iconic Photo by Manfred Schiedhelm of the Team 10 Meeting in Bonnieux in 1977, 2020
← Song Ryul, Strategic Architecture, 2024 (2018)

make no meaningful difference. In 1971, Adolfo Natalini, during a lecture at the Architectural Association (AA), captured the dilemma we face: "If design is merely an inducement to consume, then we must reject design; if architecture is merely the codifying of bourgeois model of ownership and society, then we must reject architecture; if architecture and town planning is merely the formalization of present unjust social divisions, then we must reject town planning and its cities … until all design activities are aimed towards meeting primary needs. Until then, design must disappear. We can live without architecture."[3]

Capitalism exploits the human desire for comfort and convenience. It has brought about a convenience of modern life that is truly remarkable. However, continually choosing the path of least resistance diminishes our ability to confront inevitable challenges. We often resist the very actions that are necessary for progress.[4] Convenience ties us to a system that depends on perpetual growth. On a planet with finite resources—air, water, soil, and minerals—exponential growth is, in fact, impossible and inevitably leads to environmental breakdown.

Capitalism collapses without growth, it is the core principle of this economic system. And governments measure their success by their ability to foster economic growth. When comparing real GDP growth per capita in the Western world since the Industrial Revolution with annual global CO2 emissions and material extraction, all three graphs follow the same exponential growth pattern. This demonstrates a clear correlation, revealing that decoupling GDP growth from decarbonizing our environment and preserving natural resources is inherently contradictory.

"There is no true life within a false life." Theodor Adorno's aphorism in *Minima Moralia* reflects the impossibility of achieving personal happiness in the midst of a society's catastrophic developments. Modern society affects every aspect of life; the individual cannot escape it and is willingly and unwillingly complicit. Even if we strive to live authentically, the global structure of our society makes it impossible. In simple terms, we are living a false life. To address this, we must reject the logic of economic growth and seek alternative paths.

This brings us back to the UN's definition of sustainability in the Triple Bottom Line: "Meeting the needs of the present without compromising the ability of future generations to meet their own needs." In the developed world, we have blurred the line between what we need and what is simply convenient. Much of what we think we need is actually just a matter of convenience. To ensure future generations can meet their needs, we must give up some of our conveniences without sacrificing our true needs. For humanity to survive, we must ensure a fair and equal distribution of natural resources, transcending national interests and protectionism.

Felwine Sarr, the Senegalese sociologist, musician and writer, makes this point very clear: "We are living in an occidental-ocene, not an anthropocene. It is about the occident, the West. They caused the problem but want to share the responsibility. As the anthropologist Philippe Descola identified, the West separated the world into nature and culture, the big divide. Because they saw nature as an object with resources, they took possession of nature and altered

---

3. Adolfo Natalini, *Avvicinamenti all'architettura (Quaderni di progetto)* (Pontecorboli Editore, 1996).

4. Alex Curmi, "The Big Idea: Is Convenience Making Our Lives More Difficult?," *The Guardian* (November 4, 2024).

it. Even the word environment opens up a gap, because the environment is that which surrounds the anthropos, the human being is always at the centre. Now it is an "us against them." We see them as "them," but we should accomplish a collective "we." But how do we create a collective "we" with people who don't want that? Ultimately, it's about building a mankind community, a global "we."[5]

The West—essentially, the developed world—has been placed under scrutiny. We, as the West, must act, make reparations, and change, without pointing fingers or waiting for a 'global solution.' The phrase "when you're accustomed to privilege, equality feels like oppression" highlights the core issue. For those accustomed to privilege, the idea of equality can feel like a loss or restriction, as if they are unfairly treated by having to share opportunities or benefits they once took for granted. In other words, when we are used to advantages, being treated equally can feel like a disadvantage. We must overcome this delusion and radically change our way of life to achieve global equality. A common standard for humanity cannot be sustained at the current level of the West. We must give up convenience and focus on meeting our true needs.

*

Again by coincidence of history, the Curators Collective (CC) was formed in 2020, initiated by the Korean Pavilion during the year-long postponement of the 17th Venice Architecture Biennale due to the COVID-19 pandemic. For the first time in the Biennale's history, curators and participants from 51 nations collaborated in advance, exchanging ideas and supporting one another over more than a year. The CC marks a shift of the Biennale from international competition towards collaborative strategic action.

Initially created to share information on navigating the pandemic amidst a lack of guidance from the Biennale Foundation, the CC quickly evolved into a mutual support network. Members helped each other overcome challenges like travel restrictions and logistical hurdles. This spirit of collaboration extended to content development, resulting in a three-day midissage event featuring multiple symposia and panel discussions, and an international student competition. The competition focused on repurposing leftover installation materials and the winning projects were realized within the Biennale grounds, advancing the discourse on sustainability.

The most challenging task was drafting a joint manifesto. This process revealed the immense difficulty of achieving a shared understanding of architecture's current state and outlining a unified vision for its future. First some national pavilions quit the CC when their local commissioners started to micromanage their curators which made a collaboration unfeasible for them. Then several pavilions dropped out when their individual agendas did not find a majority.

Unfortunately, the attempt to institutionalize and fund the CC after the Biennale through an EU network grant failed. Two years later, interest in the CC waned among the curators of the 18th Venice Architecture Biennale as necessity

---

5. Felwine Sarr in Marc Bauder's documentary film 'Who We Were' (2021), based on Roger Willemsen, *Who We Were: a Speech in the Future* (2016).

to collaborate post-COVID ceased and national interests reasserted dominance.

Nevertheless, this example highlights a potential future role for the Venice Architecture Biennale: not merely as an exhibition but as a platform for collaboration and consensus-building to advance architecture. Such a reimagining would require less emphasis on national pavilions and individual agendas and greater focus on fostering a common understanding of global challenges. Every nation would need a seat at the table, potentially supported by a budgetary contribution from participating nations to enable less-funded countries to join. In regard to a necessary transition from current architecture towards strategic architecture, the actual "architecture" at this point of time lies in the development of a definition of architecture fit for the challenges of the 21st century. This can only be achieved through a democratic, collective consensus within the global architectural community—not through the dominance of the discourse by a few influential nations.

*

We have come to a point where we have to admit to ourselves that all mechanisms humanity implemented to avert the self-inflicted climate crises are failing. Seven out of nine planetary boundaries, a set of limits of human activities on the Earth system beyond which the environment is not able to self-regulate anymore, have been breached. Since the 2010s, the effects of climate change have become impossible to ignore, with increasing heat waves, droughts, wildfires, extreme rainfall, flooding, rising sea levels, crop failures, and the collapse of ecosystems.

All legally binding aims of the Paris Agreement, adopted by 196 parties at the UN Climate Change Conference in 2015, are being missed. We have exceeded in the last two years the 1.5 °C limit to global warming above pre-industrial levels, with nature on a rapidly heating planet losing its ability to remove CO_2 from the atmosphere. In his book *Slow Down*, philosopher Kohei Saito convincingly debunks the strategies currently being employed to avert the climate crisis, such as the Green New Deal, green economic growth, decoupling, carbon offsetting, increased efficiency, techno-optimism, acceleration, carbon capture, geoengineering, and dematerialization, among others.[6] He concludes that only a radical degrowth approach can have any significant impact, but he questions society's willingness to transition away from the current global economic system.

In a 2024 survey by *The Guardian* among 380 leading climate scientists, 77% of respondents predict global temperatures will reach at least or more than 2.5C above pre-industrial levels with disastrous results for humanity. They stated that they had been left feeling hopeless, infuriated and scared by the failure of governments to act despite the clear scientific evidence provided.[7] It has become unequivocally clear that we are changing the Earth system in a way and a speed that has never been seen before, and that our social systems can no longer keep up with these changes. In the forthcoming decades billions of people will be pushed outside the climate niche, the period of stability of the Holocene, in which human

---

6. Kohei Saito, *Slow Down: The Degrowth Manifesto* (Astra House, 2024).

7. Damian Carrington, "Hopeless and Broken: Why the World's Top Climate Scientists are in Despair," *The Guardian* (May 8, 2024).

society developed. Resource allocation conflicts will lead to wars, billions of people will be forced to migrate, millions will die. With environmental collapse being imminent, societal collapse is inevitable.

The economist Danilo Brozovic in a recent meta-study on failed civilizations concluded that modern society requires radical transformation to ensure its survival.[8] He references Joseph Tainter's "peak complexity" theory, which posits that societies solve problems by investing resources, but as complexity grows, problems become harder and demand even greater resources. While technological innovations can temporarily simplify these challenges, this process cannot continue indefinitely, ultimately leading to collapse. Similarly, the "sunk-cost effects" theory by Janssen, Kohler, and Scheffer suggests that societies resist abandoning systems they are deeply invested in, like our current global economic system, despite bleak prospects. Other researchers attribute collapse to social hubris, arrogance that blinds societies to warning signs and prevents corrective action. The greatest challenge and most significant barrier to action lies in persuading people of the urgent need to overhaul political systems, policies, and institutions.

*

When searching for the single most effective tool to address the climate crisis, we find that the United Nations (UN) is the only mechanism we have to force a fragmented world to form consensus and take legally binding climate action. However flawed and ineffective due to its structure, which prioritizes national interests over a global common good, it is the only body that actually actively contributed to a global solution, just to mention the Paris Agreement.

As architects, perhaps we need to take a lesson from the UN: real change cannot come from individual, national, or even transnational efforts alone. Only a global consensus and framework for understanding and practicing architecture can drive meaningful progress. The 20th century, which shaped our current understanding of architecture and significantly contributed to the challenges we face today, continues to exert an outsized influence on the 21st century. It will take time to move beyond its legacy. To do so, we must reinterpret the architectural history of the 20th century, shifting away from its mainstream narratives toward alternative perspectives that address the needs of the present century—much as Modernism redefined 19th-century history to suit its goals.

The 1968 Venice Biennale was marked by anti-war and anti-establishment student protests, during which the Giardini was occupied, leading to brutal police crackdowns. In solidarity with the protests, many artists withdrew their works, resulting in a partially empty exhibition at its official opening. These events prompted a revision of the Biennale's statutes in 1973, reflecting societal changes and adopting a more provocative curatorial agenda, transforming the Biennale into a forum for cultural debate. Its relevance today is directly tied to these reforms. Perhaps it is time for similar protests to pressure the Biennale Foundation into systemic reforms that reflect contemporary needs and ensure its

---

8. Danilo Brozović, "Societal Collapse: A Literature Review," *Future*, 145, no. 3 (December, 2022).

continued relevance in the decades to come.

*

As the economist Dennis Snower points out: "We understand very well within our community that we have to put ourselves aside in order to give the community the opportunity to develop and to be of use to everyone. What we as humanity have not understood is that social groups [as in individual nations] have to take a step back in the same way in order to give the global community the opportunity to deal with the big global problems."[9]

The fact that we as a species are so successful is due to the ability to cooperate with one another. We do this with specific stories—stories that could be called identity-forming narratives. Narratives that assign social roles to people, that place them in clearly defined relationships with one another, that give them an idea of who they are and where they belong. The stories we tell each other right now are of national identity, of a global neoliberal economic system, of architecture as is, of a Biennale of "we against them." We have to change these narratives towards a global identity. We all know it but are still not fully aware.

The majority of leading climate scientists agree that we can no longer prevent a full-scale climate breakdown in time. However, this does not absolve us of the responsibility to do everything possible to mitigate its impact. Even a 0.1°C reduction in global warming could save hundreds of thousands of lives and ease the burden on millions more.

Our final chance lies in forming a collective "we" that speaks with a unified voice in the push for change. Provided we restructure and redefine its purpose accordingly, the Venice Architecture Biennale would be uniquely positioned to serve as a platform for negotiating such a consensus, and perhaps the only effective forum for this effort.

9. Dennis Snower in Marc Bauder, Ibid., 5.

TEMPORARY PERMANENT:
A GENEALOGY OF REAL-SCALE
ARCHITECTURE EXHIBITIONS

LÉA-CATHERINE SZACKA

Designed by Kim Seok Chul and Franco Mancuso, and built between 1994 and 1995, the Korean Pavilion was the last national pavilion to be built in Venice's Giardini della Biennale. Like the other pavilions of the Biennale, it proposes a particular aesthetic and spatial response that represents national identity and acts as the container for art and architecture exhibitions, while negotiating between the transient and the permanent on display.

Tucked away in a secluded spot, between the Japanese and the German pavilions, Kim and Mancuso's building is conceived as a lightweight structure of steel and glass, attached to an existing small 1930s brick building and appearing completely immersed in the nature around it. Its complex plan and undulating façade were the result of the surrounding environment made of trees and other previously built pavilions.[1] Developed off site and assembled in situ, the pavilion is indeed part of a collection of twenty-six permanent constructions erected in the Giardini della Biennale between 1887 and 1995,[2] an emblematic—and somehow anachronistic—ensemble that corresponds to a miniaturisation of the world. Together, these pavilions showcase a series of styles (from Neo-classical grandeur to neo-Palladianism and Secessionist atmosphere), materials (from brick to concrete and from wood to steel and glass), and spatial solutions (from the "alternation of closed and open spaces" to the "continuity between interior and exterior" and from the filtering of light through a counter-ceiling of brise-soleils to a space with the highest degree of flexibility)[3] that constitute and one of the most striking and internationally known and visited examples of real-scale open-air architecture exhibitions on the planet.

In the context of the Biennale, the Korean Pavilion, like all other national pavilions, has a specific purpose. Yet, it is also part of a very long history of construction covering a wide range of geographical, cultural and functional contexts which relate to a number of fundamental and ontological questions on the thin line between reality and representation, questions of permanence and temporality and the productive border between art and architecture. According to Barry Bergdoll, pavilions are "a place for the imagination" and "new forms of

1. Marco Mulazzani, *Guide to the Pavilions of the Venice Biennale Since 1887* (Electa, 1988), 26.

2. The first building in the Giardini was dessigned in 1887 by architect Raimondo D'Aronco for the National Art Exhibition of the Venice Bienale. The first national pavilion to be built was for Belgium by architect Leon Sneyers in 1907. For more on the pavilions of the Venice Biennale, see Mulazzani, *Guide to the Pavilions of the Venice Biennale Since 1887*.

3. Mulazzani, *Guide to the Pavilions of the Venice Biennale Since 1887*, 16.

sociability and innovation in architectural design" as well as "an exuberant display of a new challenge of representation and illusion."[4] Often located in gardens or parks and related to leisure activities, the pavilion was defined by English architectural historian Nikolaus Pevsner as a "lightly constructed, ornamental building, often used as a pleasure-house or summer house in a garden and also as a projecting subdivision of some larger building."[5] Not bearing the usual economic or functional constraints, they often act as a testing ground, allowing architects to explore new ideas, methods and materials. But what are these types of constructions which are neither fully permanent nor completely temporary? And what have they been in the history of architecture? Are they real pieces of architecture or tri-dimensional representation of bigger ideas? Should they be considered art or architecture, or a form of hybrid between the two? Taking the Korean Pavilion of the Giardini as a departure point, this essay explores these questions through a genealogy of real-scale architecture exhibitions.

Real-scale architecture mock-ups became important at the turn of the nineteenth century in the context of attraction parks or gardens. In 1843, Georg Carstensen opened the Tivoli Garden in Copenhagen inspired by two pleasure gardens: Paris's Jardin de Tivoli that existed from 1795 to 1842 and London's Vauxhall Garden, in activity from 1785 to 1859.[6] In the popular Danish amusement park, together with young architect Harald Conrad Stilling, Carstensen imagined a series of light pavilions, or "pleasure palaces" reposing Turkish, Arabian, Chinese, Pompeian and Russian architectural motifs. Part of the tradition of the Jardins-spectacles (spectacle gardens),[7] Tivoli's often flamboyant and ornamental pavilions were fables and sensorium pleasures for the visitors.

Parallel to that, from 1867 onwards, World's Fair or Universal Exhibitions were the occasion of erecting pavilions that represented the history, tradition and identity of nations, all concepts that were a major phenomenon in the 19th century. A form of architectural laboratory or gigantic open air architecture exhibition—for example, at the Paris 1900 Exposition, the national pavilions were spread out like an open-air museum along the Rue des Nations, between Les Invalides and the Pont de l'Alma—the Universal Exhibitions gave rise to innumerable ephemeral architectures which sometime survive time only to become architectural icons. From Joseph Paxton's Crystal Palace (London, 1851), to the Eiffel Tower (Paris, 1889) and from Mies van der Rohe's Barcelona Pavilion (1929), Alvar Aalto's Finnish Pavilion for the World Exposition in Paris (1937), Le Corbusier and Iannis Xenakis' Philips Pavilion at the World Fair in Brussels (1958), and Buckminster Fuller's Geodesic Dome for the American National Exhibition in Moscow (1959), these architectures can be considered as part of an unwritten history of 20th century architecture.

Although meant as merely ephemeral structures, many of these have been saved from demolition or reconstructed. The most famous of these architectures caught between ephemerality and permanence is the case of the German pavilion, built by architect Mies van Der Rohe for the 1929 Universal Exhibition in Barcelona. Despite being a representation of the German national identity, Mies' building was mainly an example of the type of pavilions that proliferated after 1900,

---

4. Barry Bergdoll, "The Pavilion and the Expanded Possibilities in Architecture" in *The Pavilion: Pleasure and Polemics of Architecture* (Hatje Cantz, 2009), 14.

5. J. Fleming, H. Honour and N. Pevsner, *Dictionary of Architecture and Landscape Architecture* (Penguin Books, 1999), 427. Cited in Gonca Zeynep Tunçbilek, "Temporary Architecture: The Serpentine Gallery Pavilions," Master thesis, Middle East Technical University (September 2013).

6. Mark I. West, "Tivoli Garden and Hans Christian Anderson: A Tale of Confluence," in *Storybook Worlds Made Real: Essays on the Places Inspired by Children's Narratives*, eds. Kathy Merlock Jackson and Mark I. West (McFarland, 2022), 6.

7. On this see Susan Taylor-Leduc, "Jardins-Spectacles: Spaces and Traces of Embodiment," in *Ephemeral Spectacles, Exhibition Spaces and Museums* eds. Dominique Bauer and Camilla Murgia (Amsterdam University Press, 2021), 81-106.

↗ The Glass Hall of Tivoli Gardens,
   designed by Harald Conrad Stilling
↑ National pavilions of the Paris 1900 Exposition.
→ Onyx for the 1986 reconstruction of the Barcelona Pavilion.
↘ Aerial view of Weißenhofsiedlung in 2004.

and that were "still charged with the grammar of national identity" but were mostly meant to explore the syntaxes of architecture while becoming a "locus of architectural experimentation, for unprecedented images, and new experience."[8] From the beginning, the pavilion was considered as a temporary structure for which the concept of durability—the "firmitas" of Vitruvius—was not applicable and it was dismantled after the end of the exhibition. Yet, "there was a persistent legend that the Pavilion remained in Barcelona, stored away in a secret hiding place."[9] In 1954, a quarter of century after its original construction, the re-erection of the Barcelona pavilion was first ideated by Catalan architect Oriol Bohigas. But it was not before 1980, when Bohigas became Delegate for Urban Planning that a study on the technical, philological and financial aspects of the project allowed the reconstruction to really take shape. Completed in 1986, the reconstruction of the pavilion provoked controversy, some judging it a fake or a pale copy of the original, transient building.

In more recent days, this type of small-scale construction is often linked to important cultural institutions (for example, the Venice Biennale, but also the MoMA, MoMA PS1 or the Serpentine Gallery) and is ephemeral structure and exhibition in itself rather than a container for exhibitions. In 1939, MoMA's founding director Alfred H. Barr Jr. and architecture curator John McAndrew conceived an outdoor gallery for changing exhibitions. In the following two decades, the MoMA sculpture garden hosted a dynamic range of shows, such as the presentation of life-size houses including a two-bedroom domicile designed by Marcel Breuer (1949) and a three-bedroom house designed by Gregory Ain and fitted with sliding walls that allowed for a flexible floorplan (1950) as well as a wooden house in the style of 17th-century Japanese temple architecture (1954). Part of a long tradition of mock-up, these model houses were neither reality nor representation but were located in the space in-between as they were both models in the sense of something to be copied and models in the sense of a (real) scale reproduction. Another famous case of model dwelling created in the context of exhibition is the 1927 Weißenhofsiedlung (Weissenhof Housing Settlement) built on a hill above the city of Stuttgart in occasion of the Werkbund exibition - Die Wohnung (The Dwelling) as a showcase of the most representative modern architecture.[10]

Pavilion programs are frequently realised by visual arts institutions, creating a new institutional context—located in between sculpture and architecture—for contemporary architecture.[11] One of the most known examples of these types of contemporary pavilions are produced by the Serpentine Galleries which hosts since 2000, an annual pavilion program, commissioning each year the construction of a temporary building to be erected in London's Hyde Park. A form of global "architectural experimentation", the pavilion series showcases the work of international architects and design teams for the first time in England. Quickly erected and dismantled and reconstructed in different locations (they are typically sold or pre-sold to rich client to help paying for the construction), the pavilions explore the idea of temporary architecture while, however, following a certain numbers or rules and premises. First, as pointed out by Hyejin Jung and Soram

---

8. Barry Bergdoll, "The Pavilion and the Expanded Possibilities in Architecture," Op. Cit., 19.

9. Cristian Cirici, Fernando Ramos and Ignasi de Solà-Morales i Rubio, "The Reconstruction of the Barcelona Pavilion," *ICOMOS Journal* 24 (1998), 45.

10. Matilda McQuaid, *Lilly Reich: Designer and Architect* (The Museum of Modern Art, 1996), 22.

11. John Macarthur, Susan Holden, Ashley Paine, Wouter Davidts, *Pavilion Propositions: Nine Points on an Architectural Phenomenon* (Valiz, 2018), 24.

Park, it is "a materialistic way of expressing new ideas, despite being constructed as a temporary structure in the museum's space." Second, it can be considered as a form of site-specific art project that "extends outside the white cube, making the pavilion an event-like object in the gallery garden and a folly."[12] In *Pavilion Propositions: Nine Points on an Architectural Phenomenon*, John Macarthur, Susan Holden, Ashley Paine and Wouter Davidst argue that the pavilion phenomenon has a lot to tell us about the "changing position of architecture in culture" and in particular in its relation to art.[13] They claim that being emblematic of both a form of temporary architecture and a periodicity, the pavilion often operates "between objects and buildings, or, between buildings and sculptures, as they are both on display and in use."[14]

If the Venice Biennale existed since the very end of the 19th century and has displayed architecture since the 1970s, it was only in 1991, on the occasion of its 5th International Architecture Exhibition that (under the direction of Francesco Dal Co) the institution decided to model this edition after the Art Exhibition, inviting national pavilions to participate. That year, just a few years before the construction of the Korean pavilion by Kim and Mancuso, a new conflation between architecture container and architectural content took place for the first time, as the Giardini of the Biennale, itself an open-air permanent architecture museum, became the setting for a series of national competing displays.

12. Hyejin Jung and Soram Park, "Pavilion as an Architecture of New Placeness: A Case of Serpentine Pavilion Project," *Journal of Asian Architecture and Building Engineering*, 22, Issue 1 (2023), 84-95.

13. John Macarthur, et all., *Pavilion Propositions: Nine Points on an Architectural Phenomenon* (Valiz, 2018), 11.

14. Ibid., 9.

A
CURATORIAL
FABLE

JUNG JINHO

Little Toad —
Little Toad —

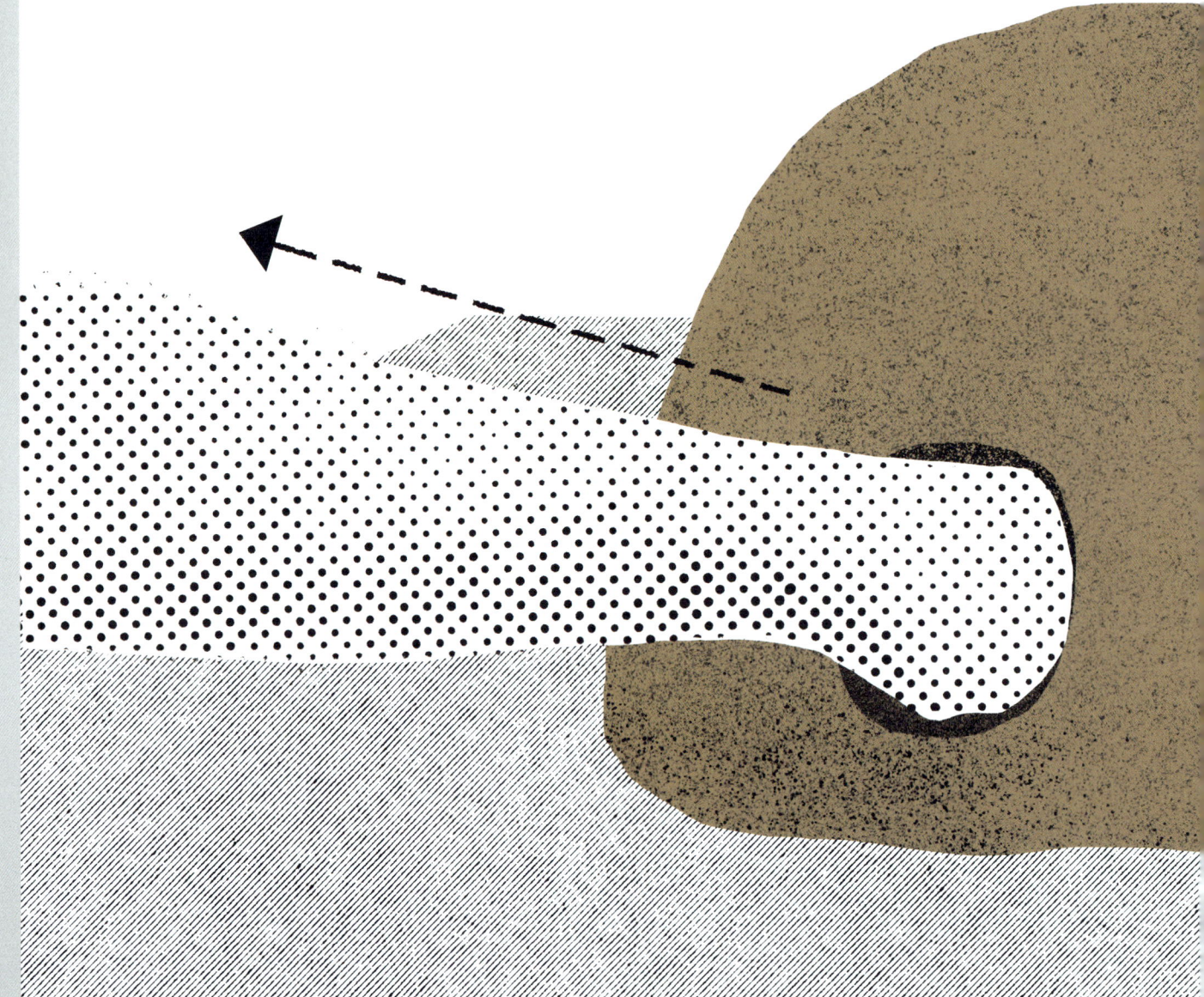

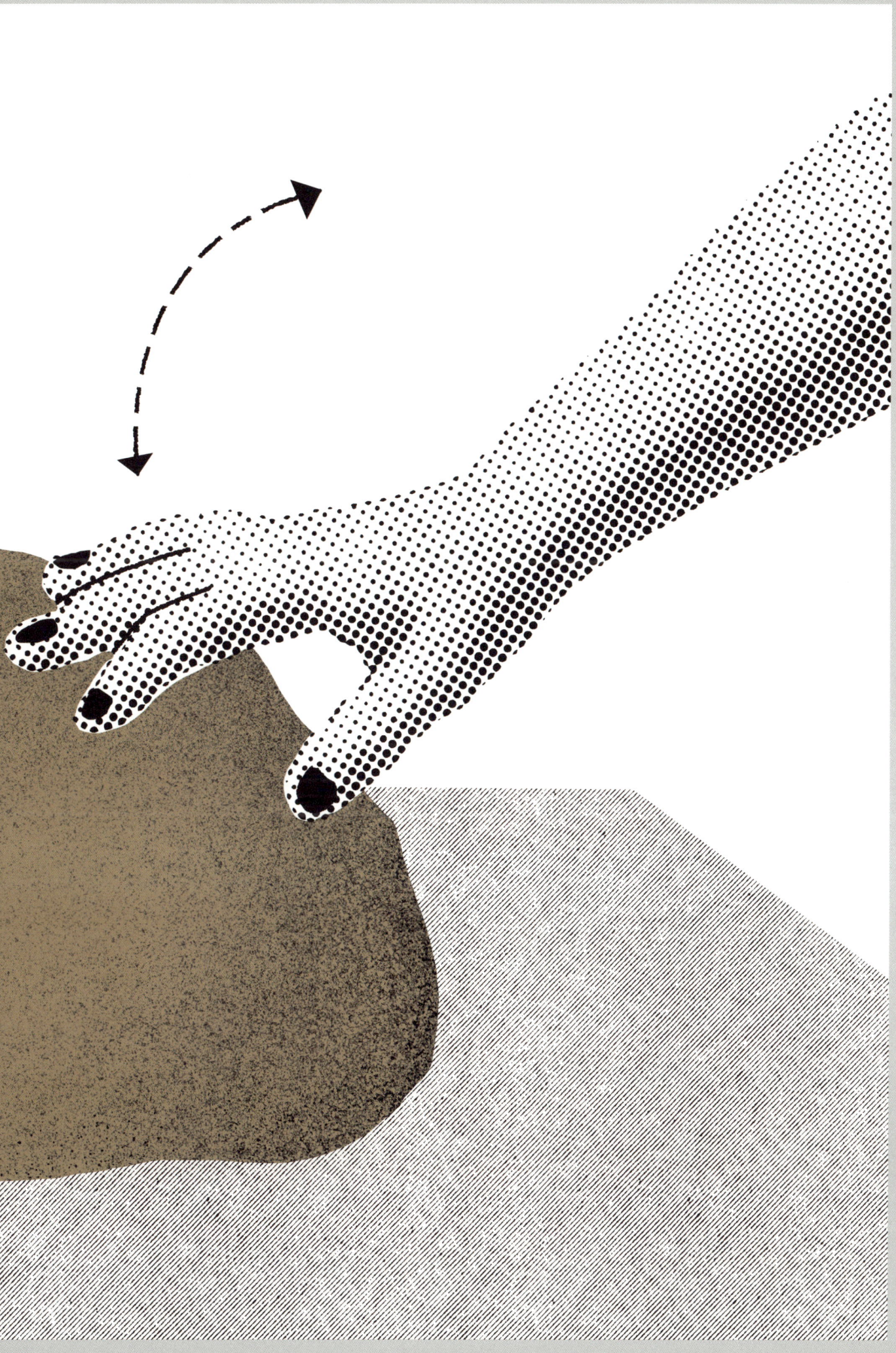

In return for
a new house

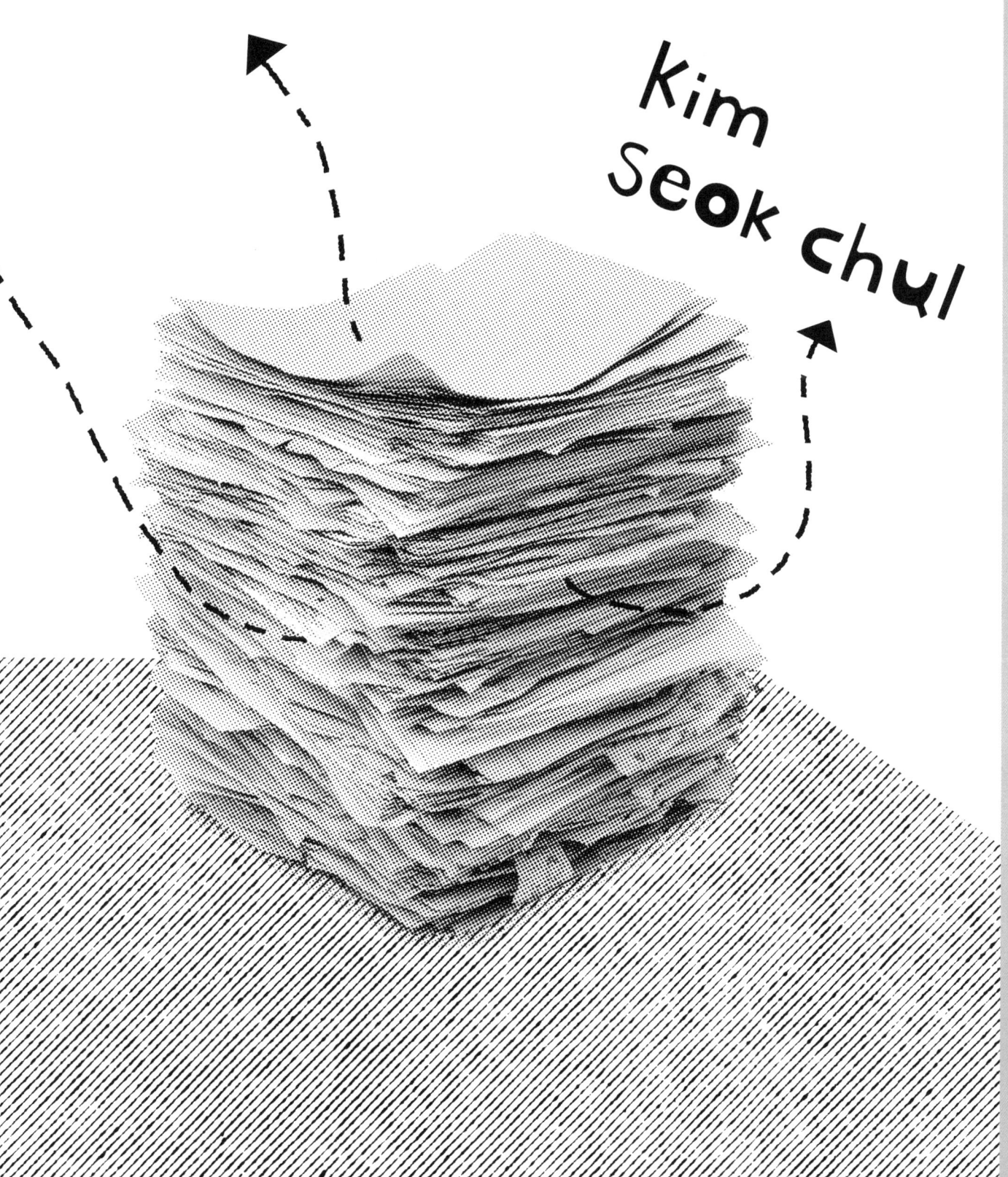

Archival materials:
©Mancuso e Serena Architetti Associati. Courtesy of ARKO Arts Archive, Arts Council Korea.
Courtesy ©Archivio Storico della Biennale di Venezia, ASAC

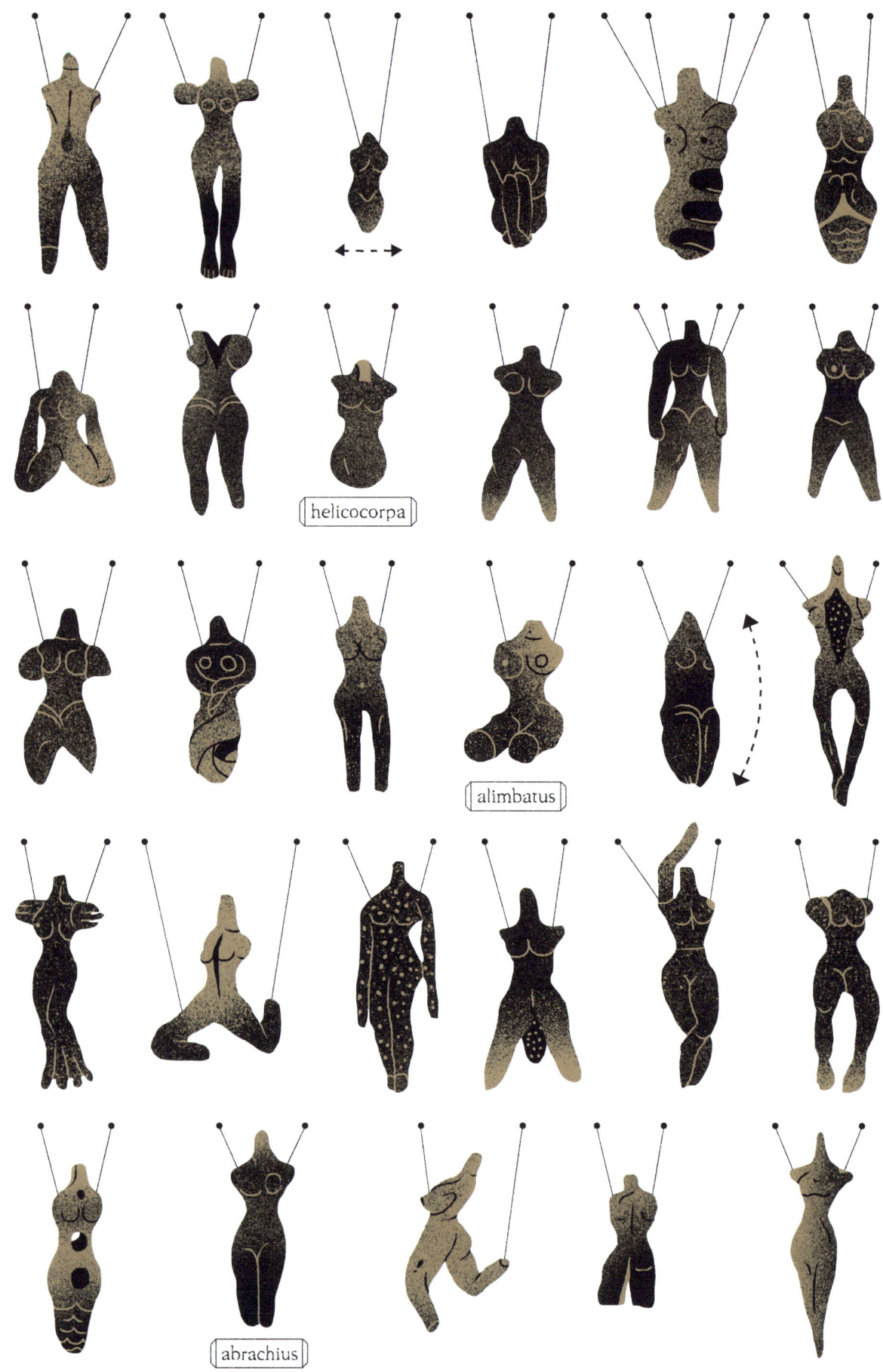

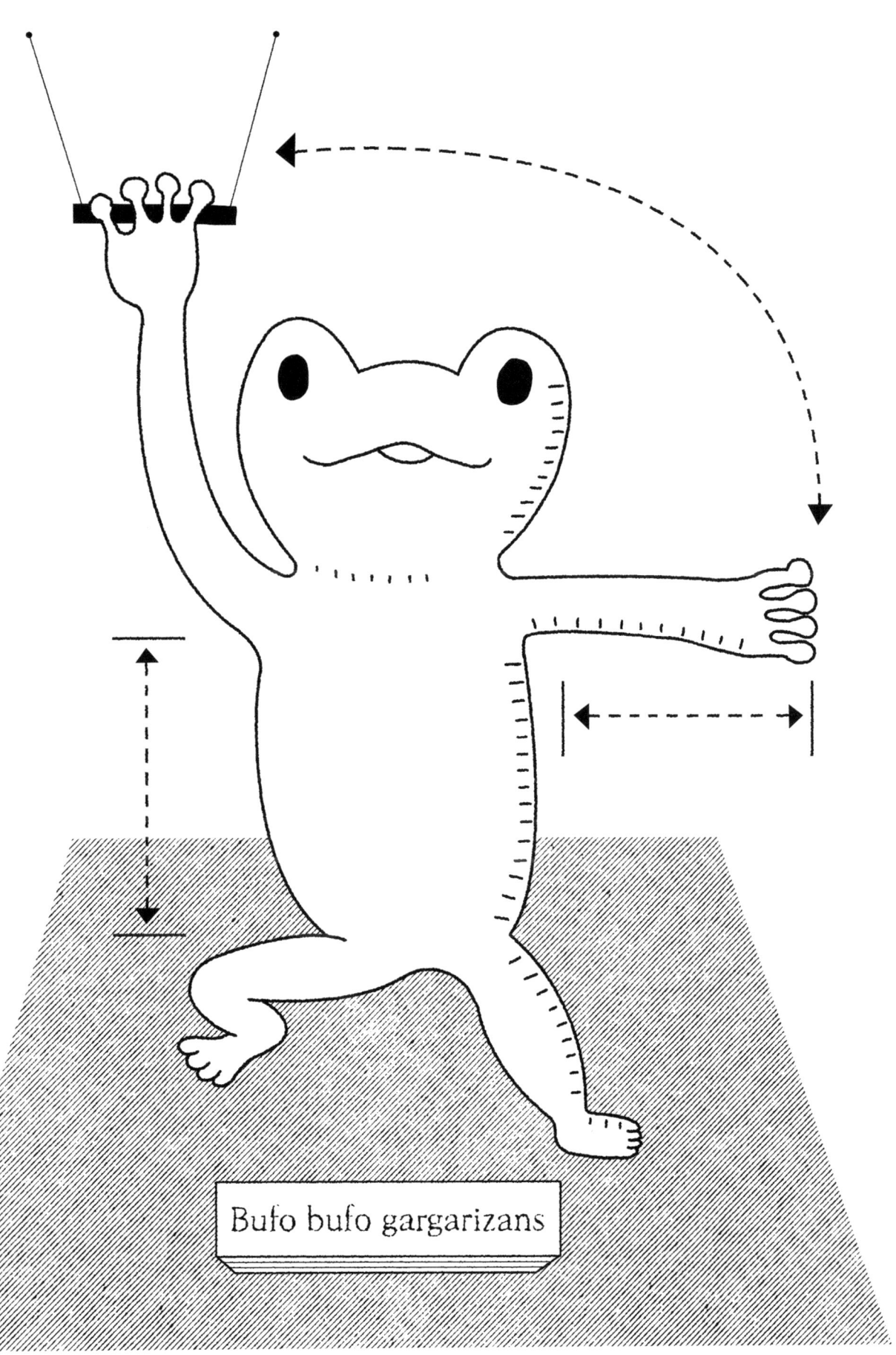

Bufo bufo gargarizans

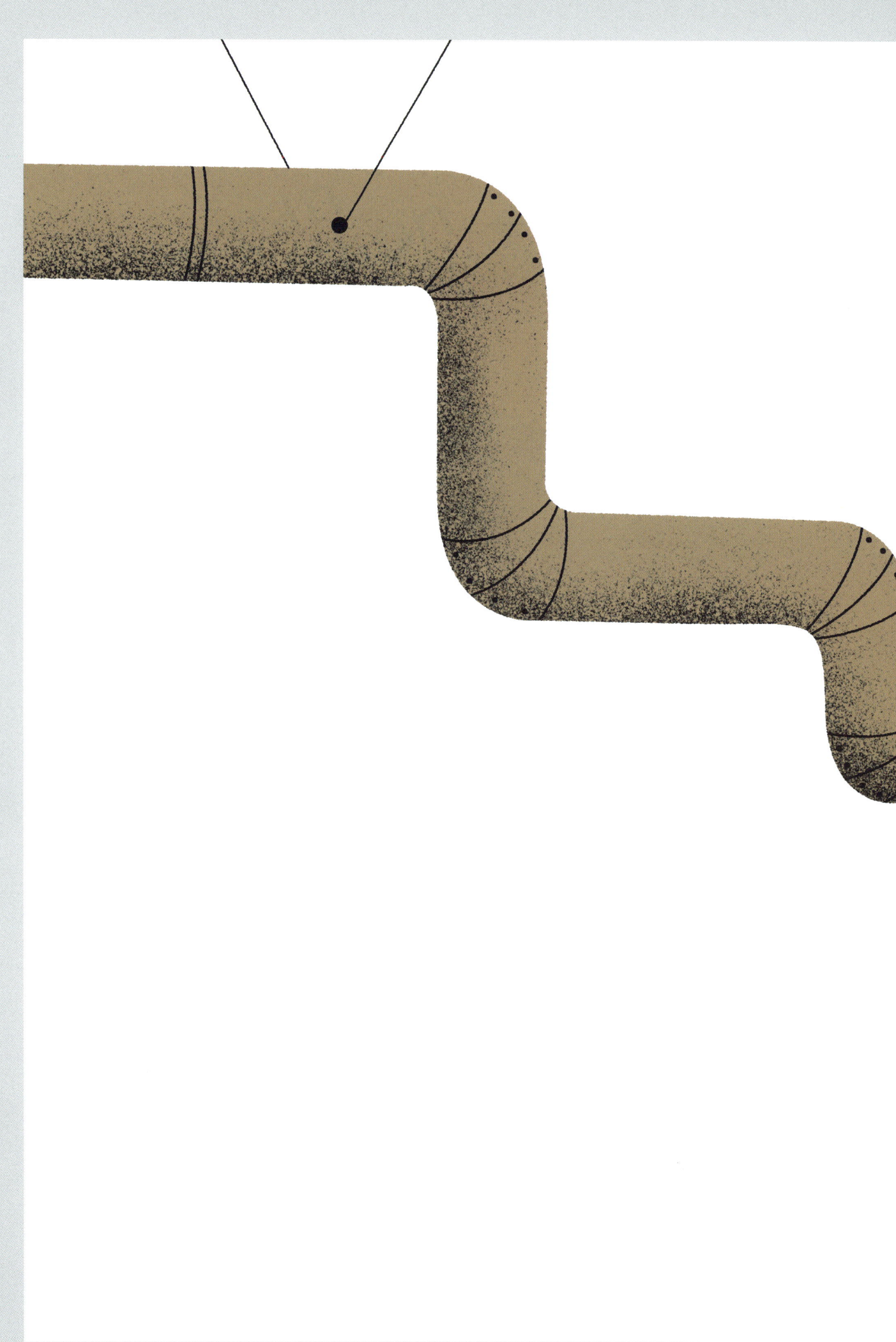

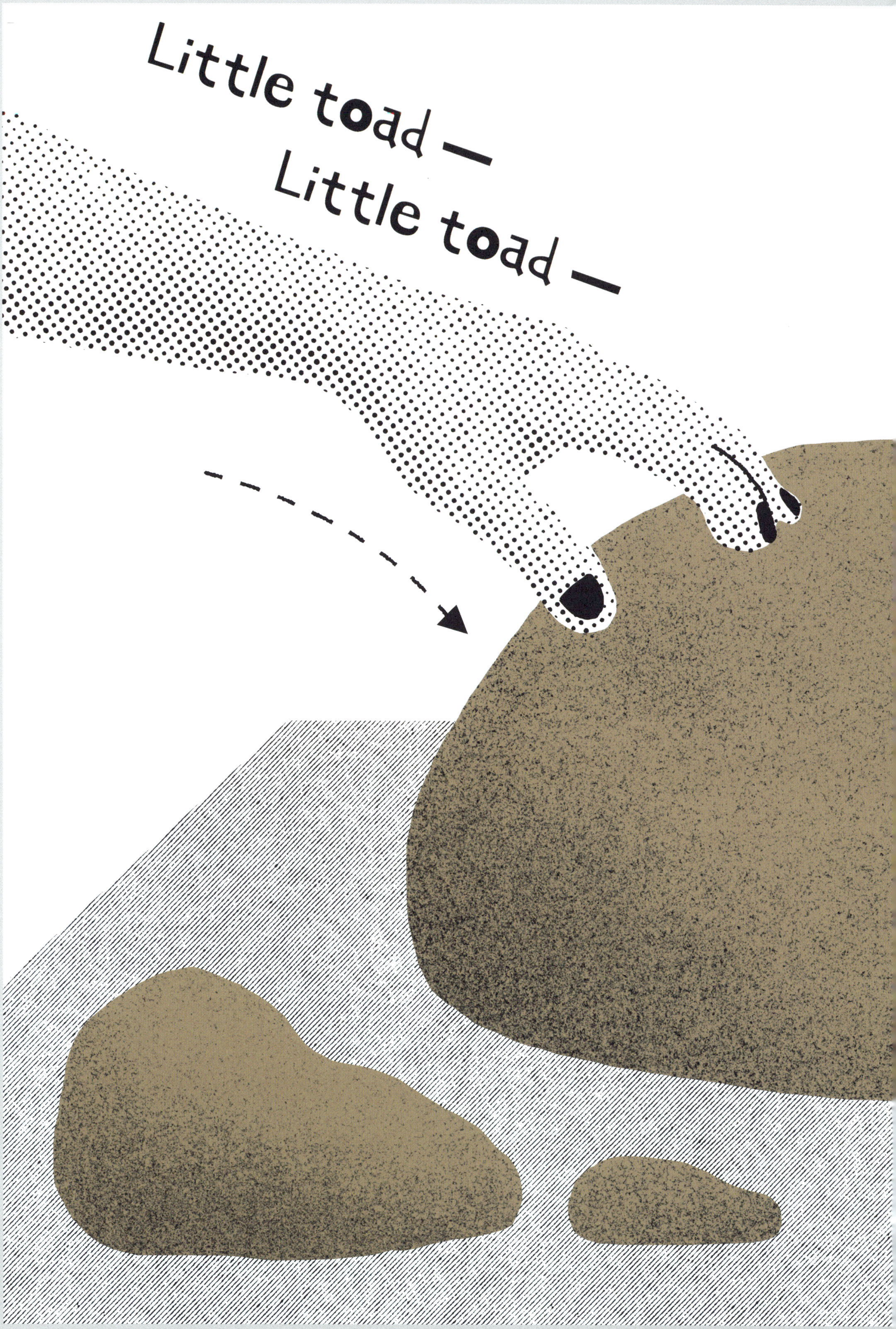

Little toad —
Little toad —

I'll give you an old house
In return for a new house

Biographies

CURATORS

CURATING ARCHITECTURE COLLECTIVE (CAC) is a curatorial group dedicated to exploring the intersections of architecture, text, objects, and space. Founded by Chung Dahyoung, Kim Heejung, and Jung Sungkyu, CAC operates as a collaborative curatorial platform engaging in diverse artistic and architectural practices. The collective has organized significant exhibitions, including *Assembly of Air* (Factory2, 2021) and *Homely Talk: Cho Byoung-soo x Choi Wook* (DDP, 2021), which examine the evolving narratives of architecture in contemporary discourse.

Through the CAC Reading Room, the collective fosters critical dialogue by hosting forums, workshops, and research-driven projects in collaboration with leading Korean and international architects, designers, and artists. Continuing its investigation into the "life of architecture," a central theme of *Little Toad, Little Toad: Unbuilding Pavilion*, CAC is committed to expanding its curatorial vision through interdisciplinary projects that challenge conventional perceptions of built environments.

CHUNG DAHYOUNG is a curator and editor whose work explores architecture, urbanism, and visual culture through research, exhibitions, and writing. From 2011 to 2024, she served as a curator at the National Museum of Modern and Contemporary Art (MMCA), Korea, organizing numerous exhibitions, including *Figurative Journal: Chung Guyon Archive* (2013), *Papers and Concrete: Modern Architecture in Korea 1987-1997* (2017), and *Performative Home: Architecture for Alternative Living* (2024). She co-curated the Korean Pavilion at the 2018 Venice Biennale of Architecture, *Spectres of the State Avant-Garde* and received the Kim Jungchul Award from the Korean Institute of Architects (KIA) in 2024. She is currently the co-director of the CAC and an adjunct professor in the Department of Industrial Design at Konkuk University.

KIM HEEJUNG is a co-director and curator at CAC. Her research focuses on the evolving roles of contemporary architects and the mediums and methodologies used to represent architecture. She served as the coordinator for the *Young Architects Program* at the National Museum of Modern and Contemporary Art (MMCA), Korea from 2015 to 2017 and was the deputy curator of the Korean Pavilion at the 2018 Venice Biennale of Architecture. From 2019 to 2024, she was a curator at Photography SeMA, where she led curatorial initiatives related to its establishment. She is also the co-author of *Pavilion, Filling the City with Emotion*.

JUNG SUNGKYU is a curator whose work explores architecture, design and visual culture. He focuses on spatial planning related to architecture, crafts, and gardening. He served as the associate curator of *Homely Talk: Cho Byoung Soo x Choi Wook* (DDP, 2021) and conducted archival research for *Olympic Effect: Korean Architecture and Design from 1980s to 1990s* (MMCA, 2020). He also served as an assistant curator of the Korean Pavilion at the 2018 Venice Biennale of Architecture. He is currently a co-director of CAC and a co-representative of TACT.

ARCHITECTS / ARTISTS

KIM HYUNJONG completed his bachelor's and master's degrees in architecture at École Spéciale d'Architecture (ESA) in Paris, France, and founded ATELIER KHJ in Seoul in 2018. He explores cultural values across various fields, including urban culture, architecture, interior design, furniture, and art, breaking boundaries while engaging in continuous research and reinterpretation based on deep interest in materials and substances. His representative works include 'Present Perfect' (2021) and 'Jumjumjumjumjum' (2021), and he has participated in numerous group exhibitions, including *Young Korean Artists 2023* (MMCA, 2023), *Seoul Maru: Public Intervention* (Seoul Hall of Urbanism & Architecture, 2022), *Stepping on the Ground, the Floor is Raised* (Arumjigi, 2020), and *Donuimun is Open* (Donuimun Museum Village, 2018). He also continues his ongoing exhibition series, *Building*.

HEECHAN PARK studied architecture in Seoul and London. He was a recipient of the Bartlett Travel Scholarship to study the works of Alvar Aalto in Finland in 2010, and was invited to exhibit his work at the Royal Academy in London in 2014. Heechan founded Studio Heech in 2018 and is carrying out projects in the fields of architecture, industrial design, fabrication, and digital interaction. He received KIA (Korean Institute of Architects)'s Building of the Year award for 'Sanyang Brewery' in 2020, the Today's Young Artist Award presented by the Minister of Culture, Sports and Tourism in 2022. Studio Heech was shortlisted for *The Architectural Review*'s Emerging Architects in 2023. His major works include 'Sanyang Brewery' (2020), 'Seoul Urban Pinball Machine' (2021), 'Four Three House' (2023), and his publications include *Aalto, Architecture & My Travels* (2020).

YOUNG YENA is a co-director of Plastique Fantastique, together with Marco Canevacci – an art duo that creates immersive spatial installations, challenging the notion of reality and pushing the boundaries of our senses. She holds a bachelor's degree in architecture from Yonsei University and a master's degree in spatial design from Aalto University. The major projects include 'RINGdeLUXE' (Nuit Blanche, 2023), 'TREES & TRACES' (Seoul Biennale of Architecture and Urbanism, 2023), '#StayOut' (Oerol Festival, 2022), 'DOUBLE HEART' (Digital Art Festival Taipei, 2020), and 'Blurry Venice' (Venice Pavilion, La Biennale di Venezia, 2019). The installation works reexamine the relationship between space and storytelling, transforming familiar environments into stages for exploration.

LEE DAMMY is an architect and designer, and the principal of the architectural firm Flora and Fauna. She studied architecture at Seoul National University and Harvard Graduate School of Design. Lee speculates on how the presence of matter animates the vitality of space and relationships through the interplay of plants, animals, objects, and buildings, while exploring the possibilities of architectural imagery. In doing so, she revisits the landscape of industrialized and institutionalized architecture through the lenses of new nature, gender, and ornamentation. Notable works include 'Noise Center' (2023) and 'Pillar Suit' (2021), and she has participated in exhibitions including *Young Korean Artists 2023* (MMCA, 2023). Lee is a founding member of the collective Yeojiphap and co-author of *Building Role Models: Architecture Spoken by Women*. Currently, she teaches at Seoul National University.

ASSISTANT CURATOR / CATALOG EDITOR

KWAK SEUNG-CHAN is a researcher in architectural history, theory, and criticism, with a focus on alternative approaches on history-narrating of modern and contemporary Korean architecture. After receiving his Bachelor of Architecture from Korea University Department of Architecture with Lee Jeong Deok Architecture Award, he is continuing his research at Archistory KU of Korea University Graduate School. He translated several articles and books on art and architecture, and has worked in the Archive Team of Junglim Architecture (2023-2025) where he led the Hyundai Motor Company's Architectural Heritage Archiving project.

SPATIAL DESIGNER

KIM GISEOK founded Spatial Semiology in 2019, establishing a practice centered on composing spaces through reconfigurable architectural elements. His work spans commercial interiors, scenography, furniture, and objects, embracing a diverse range of sculptural forms. Currently based in Antwerp, Belgium, Giseok is focusing on object-based work beginning with his solo exhibition *Proceed* (COUR, 2024) while simultaneously pursuing the INSIDE Master's program at the Royal Academy of Art, The Hague, where he integrates academic research with spatial practice.

GRAPHIC DESIGNER

KIM YUNA worked as an in-house graphic designer at the National Museum of Modern and Contemporary Art (MMCA), Korea from 2014 to 2020. In 2020, she established a graphic studio, yunakimc. Her work spans print media, exhibition graphics, and branding. She has also served as an experimental graphic design instructor at PaTI. Recently, she has worked on exhibition identity design for *Hyundai Blue Prize Design 2022* (Hyundai Motorstudio Busan, 2023), *Moments in Serpentine Pavilions 2000-2024* (Seoul Hall of Urbanism & Architecture, 2024), and *K-Royal Culture Festival* (Korea Heritage Agency, 2023-2024).

VIDEOGRAPHER

BAEK YUNSUK studied advertising, media, and visual anthropology before founding the web documentary channel The Docent. He has produced exhibition videos using various media, including archives, interviews, and documentaries, for institutions such as the National Museum of Modern and Contemporary Art (MMCA), Korea, Gwangju Biennale, and the Korean Pavilion at the Venice Biennale. He was the producer of the feature documentary 'More' (2022) and has directed music videos for artists such as Lang Lee, Se So Neon, and Silica Gel, as well as commercials for brands. He is also active as a media artist.

ILLUSTRATOR

JUNG JINHO is a picture book artist, author, and illustrator based in Seoul. He studied architecture in university but now builds homes within the pages of picture books. He has received multiple Bologna Ragazzi Awards, including the OPERA PRIMA and Art & Architecture Design categories. His works have been published in South Korea, the United States, France, Belgium, China, and Taiwan. Some of his most notable books include *LOOK UP*, *THE WALL*, *3 Seconds*, and *The Nine-Tailed Fox*.

CATALOG EDITOR

PARK JUNGHYUN is an architectural critic and the editor-in-chief of an architectural journal, *Labyrinths*. He received his doctorate from the Department of Architecture, University of Seoul. He has produced several publications, including *Modern Architecture in the Developmental State of Korea*, and is the co-author of *Korean Architecture in the Transitional Period*, *4.3 Group, Experiment of Architopia*, and *Design Culture in the Middle Class Age*. He translated Pai Hyungmin's *Portfolio and Diagram* and John Summerson's *The Classical Language of Architecture* into Korean, and has participated as a curator of the Korean Pavilion at the 2018 Venice Biennale of Architecture, *Spectres of the State Avant-Garde*. He is currently an adjunct professor in the Department of Architectural Engineering at Yonsei University.

CATALOG DESIGNER

O HEZIN is a graphic designer who runs OYE. She has worked on several commissioned projects while simultaneously engaging in initiative projects that explore the intersection of exhibitions and publications. She was invited to the Residence programme at the OTIS College of Art and Design (2018), and the experimental workshop using the Riso stencil printing technique, Magical Riso (Van Eyck, 2016). Major exhibitions she has participated in include *Poster Show* (Likely General, 2018), *Typojanchi* (Culture Station Seoul 284, 2019), *2021 Seoul Biennale of Architecture and Urbanism*, *Young Korean Artists 2023* (MMCA, 2023), *POST/NO/BILLS #5 BHLNTTTX* (Amsterdam Museum of Urban Life, 2024), among others. In 2023, she was selected as a member of the Alliance Graphique Internationale.

RESEARCHER

LEE JUNGWON is interested in art history and design, focusing on modern and contemporary Korean art. Her research explores visual arts that have emerged through cultural exchange. She conducted research on the Lee Kun-hee Collection in National Museum of Modern and Contemporary Art (MMCA), Korea and worked as an assistant curator at Art Sonje Center. Currently, she works as a curatorial assistant at MMCA, where she coordinates international exchange and archive exhibitions.

WRITERS

CHUN JINYOUNG is a licensed architect in both Korea and Italy and a Professor of Architecture at the College of Architecture, Myongji University. He earned his bachelor's degree from Hanyang University and obtained a Dottore in Architettura from Sapienza Università di Roma. He led the project *Study for the Organization and*

Collection Development of Architectural Records of the Korean Pavilion at the Venice Biennale (Arts Council Korea, 2025), where he organized the Franco Mancuso Donated Archives and cataloged its key records.

ALICE S. KIM received her MA/PhD in Rhetoric from UC Berkeley, and is a researcher and translator living in Seoul. Her dissertation, *Airport Modern: The Space Between International Departures and Arrivals in Modern Korean National Imaginings* (2013), explores the genealogy of the postcolonial modern Kimpo airport and air travel through the lens of the contradictions of postwar SK industrialization and development. Her publications include "The 'Vietnamese' Skirt and Other Wartime Myths" in *The Vietnam War in the Pacific World* (UNC Press, 2022), "Left Out: 'People's Solidarity for Social Progress' and the Evolution of South Korean Minjung After Authoritarianism," in *From Democracy to Civil Society: The Evolution of Korean Social Movements* (Routledge, 2011), and *Globalization and Art* (co-editor, Penn State Press, 2011).

SONG RYUL is an architect, educator and publisher living and working in Seoul, Korea. She is co-principal of SUPA Song Schweitzer and editor of *SUPTEXT* magazine, where she explores fundamental issues of everyday life expressed through art and design. Her work is focused on a conceptual approach towards design, finding new ways to expand the vocabulary of architecture. www.suparc.net

CHRISTIAN SCHWEITZER is an architect, writer and educator living and working in Seoul, Korea. He is co-principal of SUPA Song Schweitzer and co-founder of the Ernst May Museum in Frankfurt, Germany. He is working within the narrow intersection of conceptual design, art and architecture theory with special emphasis on understanding and transforming the contemporary city through its specific sociocultural context. www.suparc.net

LÉA-CATHERINE SZACKA is Senior lecturer (Associate Professor) in Architectural Studies at the University of Manchester and member of the Manchester Architecture Research Group (MARG). Her work focuses on the history of architecture exhibitions and the history and theory of postmodern architecture, including in its interaction with media and the rise of environmental awareness.

**Little Toad, Little Toad:
Unbuilding Pavilion**

The Korean Pavilion
19th International Architecture Exhibition -
La Biennale di Venezia
May 10 - November 23, 2025

Commissioner

Curators
CAC
(Chung Dahyoung,
Kim Heejung,
Jung Sungkyu)

Architects/Artists
Kim Hyunjong
Heechan Park
Young Yena
Lee Dammy

Assistant Curator
Kwak Seung-Chan

**On-site Manager/
Construction Supervisor**
Kim Eun Jeong

Spatial Design
Kim Giseok

Graphic Design
Kim Yuna

Videography
Baek Yunsuk

Illustration
Jung Jinho

Research
Lee Jungwon

English Copy Editor
Alice S. Kim

Transportation
ARTrans

PR
MAG PR & Image

Web Developer
DEERSTEP

Arts Council Korea
Byoung Gug Choung
(Chairperson), Song Si
Kyeong (Director of General
Secretariat), Jade Keunhye
Lim, Hyeju Choi, Sunhee
Yeo, Ji Yeon Yu, Haebin Lee
(ARKO Art Center),
Choi Jung Eun (Arts
Advocacy Center)

Partner
IKEA Korea

Sponsors
Samsung Foundation of Culture
JOOSUNG DESIGNLAB
JUNGLIM ARCHITECTURE
PNL Co., Ltd. / KIM, SEOK WOO
HAEAHN Architecture, Inc
Gansam Co.,Ltd
SPACE GROUP
THE_SYSTEM LAB
DUOMO
J.archiv

Technical Support
LG OLED AI
LG StanbyME

Scenography Supported by
VOLA

Collaborator
LUSH

Supporters
BCHO Architects Associates
ONE O ONE architects
Harper's BAZAAR Korea
WOOYOUNGMI

Artists Supported by
STRX/UPPERHOUSE
Luna&Company
KM Beam
Helinox

Special Thanks to
ARKO Arts Archive
Shin Haewon
Chun Jinyoung
Cho Byoungsoo
Choi Wook
Franco Mancuso

Associated Institution
Seoul Metropolitan Government
(Seoul Biennale of Architecture and Urbanism)

This publication is the official catalog of the exhibition:

**Little Toad, Little Toad:
Unbuilding Pavilion**

The Korean Pavilion
19th International Architecture Exhibition -
La Biennale di Venezia
May 10 - November 23, 2025

The Korean-language version of the text featured in this catalog can be accessed via the QR code.

본 도록에 실린 글의 한국어판을 QR 코드를 통해 확인할 수 있습니다.

Editors
Park Junghyun, Kwak Seung-Chan

Designer
O Hezin

Contributors
Chun Jinyoung, Chung Dahyoung, Jung Jinho, Jung Sungkyu, Alice S. Kim, Kim Giseok, Kim Heejung, Kim Hyunjong, Kwak Seung-Chan, Lee Dammy, Heechan Park, Christian Schweitzer, Song Ryul, Léa-Catherine Szacka, Young Yena

English Copy Editor
Alice S. Kim

Publisher
Propaganda

Paper Support
Hansol Paper

Hansol Hi-Q Millennium Art 250g/m²
Hansol Cloud 80g/m²
Hansol Newplus 70g/m², 120g/m²

First published in April, 2025
Printed by Kumkang Printing in Paju, Korea

© Propaganda, The Korean Pavilion 2025,
Arts Council Korea, and the contributors

All rights reserved. No part of this publication may be reproduced or transmitted in any form or by any means, electronic or mechanical, including photocopy, recording, or any other information storage and retrieval system, without prior written permission from the copyright holders.

ISBN 978-89-98143-91-6

www.korean-pavilion.or.kr